ELECTRICIAN'S
INSTANT ANSWERS

ELECTRICIAN'S INSTANT ANSWERS

David Tuck
Gary Tuck

McGRAW-HILL

New York   Chicago   San Francisco   Lisbon   London
Madrid   Mexico City   Milan   New Delhi   San Juan
Seoul   Singapore   Sydney   Toronto

The McGraw·Hill Companies

Cataloging-in-Publication Data is on file with the Library of Congress

1 2 3 4 5 6 7 8 9 0 DOC/DOC 0 9 8 7 6 5 4 3

ISBN 0-07-140203-9

The sponsoring editor for this book was Larry S. Hager and the production supervisor was Sherri Souffrance. It was set in Stone Sans by Lone Wolf Enterprises, Ltd.

Printed and bound by RR Donnelley.

McGraw-Hill books are available at special quantity discounts to use as premiums and sales promotions, or for use in corporate training programs. For more information, please write to the Director of Special Sales, McGraw-Hill Professional, Two Penn Plaza, New York, NY 10121-2298. Or contact your local bookstore.

CONTENTS

ABOUT THE AUTHORS

David and Gary Tuck are a father and son team who have devoted their lives to the electrical trade. Both men are master electricians and operate a family-owned business in Fort Fairfield, Maine. Over the years, the Tuck team has done residential, light commercial, commercial, and industrial work. Their services have been in such high demand that they have traveled hundreds of miles to take on large projects. They are known for responding to the most basic requests from area homeowners. Having been in business for decades, the Tuck name is known to nearly everyone in their community. Their well rounded and deep rooted experience brings this book to life with years and years of hard earned, hands on knowledge.

UNDERSTANDING CIRCUITS

U nderstanding circuits is a necessary part of an electrician's education. There are many types of circuits in use. Every type of circuit has its own job to do. Choosing the right circuit for a job is essential to a successful electrical installation. Are you familiar with all of the major types of circuits? If not, you will be more informed once you finish reading this chapter. The following circuits listed are the ones encountered most often. Keeping these circuits fresh in your mind can prove helpful in the field.

SERIES CIRCUIT

A series circuit is one where all devices on the circuit are connected one after another (Figure 1.1). Every device on this type of circuit receives the same amount of power. Streetlights are often wired on a series circuit.

PARALLEL CIRCUITS

Parallel circuits are sometimes called multiple circuits or shunt circuits. These circuits are similar to series circuits, but not exactly the same (Figure 1.2). Parallel circuits have all devices arranged so that

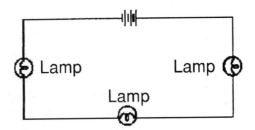

FIGURE 1.1   Diagram of a series circuit.

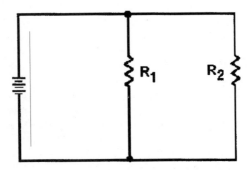

FIGURE 1.2   Diagram of a simple parallel circuit.

current is divided between them. How do parallel circuits and series circuits differ? Series circuits maintain constant power to devices connected to them and the generated electron moving force (emf) varies with the load. In a parallel circuit, the current running through the generator varies with the load and the generator emf is maintained practically constant.

PARALLEL-SERIES CIRCUIT

A parallel-series circuit is made up of many minor circuits in series with each other and with several of these series circuits then connected in parallel (Figure 1.3).

fastfacts

White wires are generally neutral wires. Neutral wires are current carrying conductors that are grounded at the service entrance point. Although there is no voltage present on these conductors they do carry current. Other acceptable colors for neutral wires are gray or three continuous white stripes on other than green insulation.

Black wires are hot. They carry power from a service panel to a device. These conductors are ungrounded and carry voltage as well as current. Hot wires can be virtually any color other than white, gray, green or three continuous white stripes on other than green insulation. Red wire, and other colored wire, is usually hot. The color coding of these hot wires can be used to identify which circuits they are serving.

White wire that has black electrical tape around the wire insulation is hot. You have to be careful not to confuse a plain white wire with a white wire that has been designated as a hot wire. This method is acceptable when using multi-conductor cables. When using single conductor wiring methods, conductors in size six American wire gauge and smaller must have continuous installation of the proper color code. No taping is allowed on these sizes.

Green wire is a ground wire. Bare wire is also a ground wire. A green wire with one or more yellow stripes is also a ground wire. A ground wire is a conductor that carries current only in the event of a ground fault.

 Don't Do This! Use green wire nuts only on ground wires. Never put hot wires together with green wire nuts.

SERIES-PARALLEL CIRCUIT

A series-parallel circuit consists of many minor circuits being connected in parallel and then several of the parallel-connected minor circuits are connected in a series across a source of emf. You won't see many of these circuits because they are rarely used (Figure 1.4).

DIVIDED CIRCUIT

A divided circuit is little more than a form of a parallel circuit. When talking about a divided circuit, the distinction is that the divided circuit is an isolated group of a few conductors in parallel, rather than a group of a large number of conductors in parallel.

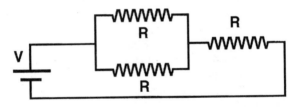

FIGURE 1.3   Diagram of a simple parallel-series circuit.

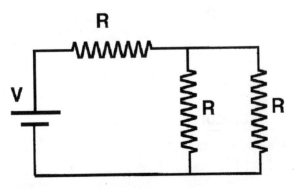

FIGURE 1.4   Diagram of simple series-parallel circuit.

MULTIPLE CIRCUITS

Multiple circuits are used for the distribution of electrical energy for lighting and most other power work. However, when streetlights are being wired, they are wired with a series circuit.

FEEDER CIRCUITS

Feeder circuits consist of a set of conductors in a distributing system that extends from an original source of energy, such as a service panel, to a distributing center without having other circuits connected to it between the source and the center. A good example of a feeder circuit is the heavy cable that runs from the main panel board in a house to the subpanel in the garage.

SUBFEEDER CIRCUITS

Subfeeder circuits are an extension of a feeder circuit. These circuits are fed through a cutout or a feeder or another subfeeder and they do not have other circuits connected to them between the two distributing centers.

MAIN CIRCUITS

A main circuit is any circuit to which other energy-consuming circuits are connected through automatic cutouts at different points along its length, which are of the same size of wire for its entire length, and which have no cutouts in series with it for its entire length. Energy-using devices are never connected directly to a main circuit. Cutouts are required between main circuits and devices that require electric power.

SUBMAIN CIRCUITS

A submain is a subsidiary main that is fed through a cutout from a main or some other submain. Branch circuits or services can be connected to a submain through cutouts. Wire used for a submain is usually smaller than the main or submain which supplies the submain.

TAP CIRCUITS

Tap circuits are created when a single energy-using device is fed from a circuit that is connected directly to the tap circuit without the interposition of a cutout.

BRANCH CIRCUITS

A branch circuit is the circuit conductor between the final circuit breaker and electrical outlet. Branch circuits are the most common type of circuits in a building (Figure 1.5). Branch circuits supply power for:

- *Lighting*
- *Appliances*
- *Convenience receptacles*
- *Motors*
- *HVAC equipment*
- *Smoke detectors*

Typical branch circuits required in a house are:

- *Small appliance branch circuits: two or more 20 amp branch circuits must be installed for receptacles in the kitchen, dining room, or similar areas. Generally speaking, no other receptacles are allowed on these circuits.*
- *Laundry branch circuit: At least one 20 amp circuit must be installed for a laundry receptacle. No other outlets are allowed on this circuit.*
- *Bathroom branch circuit: At least one 20 amp circuit must be installed for the bathroom receptacles. Other equipment in the bathroom such as lights and exhaust fans may be connected to this circuit. No outlets outside the bathroom are allowed on this circuit.*
- *Arc-fault circuit: All branch circuits supplying power to bedrooms must be protected by arc fault circuit interrupters. Arc fault circuit breakers fit in a standard circuit breaker panel but have the neutral wire connected to them as well as the hot wire. In some localities smoke detectors are exempt from this requirement.*
- *Dedicated branch circuits: A dedicated or individual branch circuit is one that supplies power to a specific piece of equipment.*

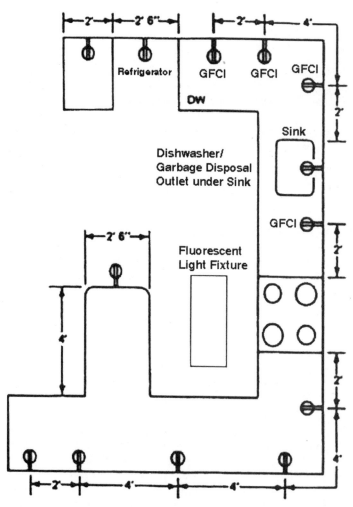

FIGURE 1.5 Diagram of typical kitchen appliance branch circuits.

- *Dryer circuit: This circuit consists of four #10 wires. Black and red are hot, white is neutral and green or bare is ground. The two hot wires are connected to a 30 amp 2 pole circuit breaker. White goes on the neutral bar and green goes on the ground bar. The other end of the circuit is connected to a 4 wire dryer receptacle.*

fastfacts

When more than one receptacle on the same yoke is supplied by more than one branch circuit, there must be a means provided at the panel box where the branch circuits originate to simultaneously disconnect the ungrounded conductors supplying those receptacles. This can be accomplished with a double pole circuit breaker or by installing a handle tie on two single pole circuit breakers.

- *Range circuit: This circuit consists of four #8 copper or #6 aluminum conductors. The hot conductors, colored black and red, connect to a 40 or 50 amp 2 pole circuit breaker. White goes on the neutral bar and green goes on the ground bar. The other end of the circuit is connected to a 4 wire range receptacle.*

- *Electric water heater: Electric water heaters are supplied by three #10 conductors. The two hot conductors are connected to a 25 amp 2 pole circuit breaker. The green or bare conductor is connected to the ground bar. The other end of the circuit is connected directly to the water heater. Check the wiring diagram with the water heater and connect the two hot wires and the ground to the proper points.*

- *Oil fired water heater: Oil fired water heaters are supplied with three #12 conductors. The black wire is connected to a 15 or 20 amp circuit breaker. The white wire is connected to the neutral bar and the ground wire is connected to the ground bar. The circuit should leave the panel and go to the emergency switch located at the entrance to the area containing the water heater. From there, the circuit goes to the thermal cutout located above the water heater. Circuit conductors then drop down into a service switch mounted on the water heater and from the service switch into the terminal box on the water heater. The conductors are protected by flexible conduit where they drop down from the thermal cutout located on the ceiling to the water heater.*

- *Dishwasher: The dishwasher is usually supplied by a 120 volt 15 or 20 amp circuit. The line runs from the circuit breaker panel directly to a terminal box located at the bottom of the dishwasher. Instructions supplied with the dishwasher will show you where to run the supply cable up through the floor. Leave 3 or 4 feet of slack cable* underneath the dishwasher so that it can be pulled out for servicing without disconnecting the power supply.

- Water pump: The water pump is usually supplied by a 240 volt 15 or 20 amp circuit. The supply cable consists of two #12 hot conductors and one number 12 ground conductor. Connect the two hot wires to a 2 pole circuit breaker and the ground conductor to the ground bar. The other end of the circuit goes to a service switch located near the pump. The switch must be a 2 pole unit so that it cuts all power to the pump when in the off position. The circuit from the switch then goes to the pressure switch that controls the pump.

- Oil-fired furnace or boiler: These units are wired the same way as an oil-fired water heater. They are supplied by a 120 volt 15 or 20 amp circuit. The circuit must go through the safety switch, thermal cutout, and service switch before connecting to the terminal box on the oil-fired unit.

- Split-wired branch circuit: This is where two branch circuits supply power to each plug on a duplex receptacle. This circuit is supplied with four number 12 conductors. Black and red are hot and connect to a 2 pole 15 or 20 amp circuit breaker. White connects to the neutral bar and green or bare connects to the ground bar. The circuit runs from the panel to the first duplex receptacle in the circuit. The jumper tab must be removed from the hot side of the receptacle. Red connects to one brass screw and black connects to the other brass screw. The neutral wires must be pigtailed with the pigtail connected to the neutral side of the receptacle. Grounds must also be pigtailed with the pigtail connected to the ground screw. This circuit works well over the kitchen counter top where appliances drawing heavy current are utilized.

fastfacts

Arc fault circuit interrupter devices are required in all branch circuits that supply 125 volt, single-phase, 15 or 20 amp receptacles in bedrooms of homes. Lighting outlets and switches must also be protected by arc fault circuit interrupters. Under certain circumstances smoke detector circuits are exempt from this requirement.

2

TEMPORARY WIRING

Temporary wiring is used for power and lighting in the construction of new buildings, houses, sheds, garages, and barns. Temporary wiring is also used in several other types of new construction and remodeling projects. The requirements for all wiring installations, temporary or permanent, are the same. Temporary wiring should be installed so it is not exposed to physical damage. Type NM and NMC cable (Romex) can be used in any dwelling, building, or structure without any height limitation.

Single insulated conductors are permitted where installed for the purpose of lighting (general illumination) and all lamps are protected from accidental contact or breakage by suitable lamp holders and guards and located where only accessible by qualified personel familiar with the construction and operation of the equipment and the hazards involved.

All branch circuits must be wired from an approved panel board or power outlet. All the receptacles have to be of the grounded type. All of the receptacles should be connected to the continuous grounding conductor. Temporary receptacles wired on construction sites cannot be wired to the branch circuits that supply the power for temporary lighting. All 125 volt single-phase 15, 20, and 30 amp receptacles on construction sites shall have *ground fault circuit interrupter* (GFCI) protection. Each temporary circuit installed must have a disconnecting means such as a circuit breaker. Disconnects must disconnect all of the ungrounded conductors of each temporary circuit.

On multi-wire temporary branch circuits, such as 12/3 Romex for example, all of the ungrounded circuits must be provided with a means to disconnect them simultaneously.

Temporary wiring cords and cables should be protected from accidental damage. Sharp corners, doorways, and projections must be avoided when the cords and cables pass through the building. Other places on the job where they can get pinched or cut need to be protected. A good way to protect NM cables in these areas is with short sections of flexible conduit.

Temporary cables and cords should be run in a manner so they are protected from physical and accidental damage. The cables and cords should be supported in place to ensure that they will be protected from damage. The supporting means can be staples, cable ties, tape, straps, or similar types of hangers. These supports should be used so no damage occurs to the cables that are used for the temporary power and light circuits. The temporary power and lighting circuits need to be removed as soon as the permanent wiring has been energized in the areas where the temporary wiring was done.

Temporary service requirements may vary from state to state and one jurisdiction to another. Before starting any work to install a temporary service, the power company must be consulted. The local code enforcement officer must also be contacted before any work is started. Often a temporary service panel board is mounted on a newly set pole that will be used as part of the permanent power distribution for the building being constructed. Other times, a temporary pole must be constructed.

The most common type of temporary power pole constructed on site is the 6 x 6 inch pole. In this application, three 2 x 6 inch planks are spiked together. This temporary pole must be set at least 4 feet in the ground. It should be braced by 2 x 4 inch braces in three directions, spiked to 2 x 4 inch stakes driven at least 3 feet in the

fastfacts

On construction sites, temporary multi-wiring circuit conductors do not require a splice or junction box. All splices have to be joined with a splicing device. Twist the wires together and cover the ends with an insulating device such as a wire nut or tape. The covering must be equivalent to the insulation on the conductors.

ground. This type of temporary service power pole is generally used in new home construction, but can be used in other types of construction jobs (Figure 2.1).

On some larger construction sites, the general contractors will set their own black jack poles for temporary power panels to be mounted on. On some work sites the temporary service may be installed on a tree or on the side of the construction trailer if the installation is first approved by the local district supervisor of the power company. If the man-made temporary pole is of the 6 x 6 inch construction, then certain requirements must be met. The pole must be set at least 4 feet in the ground and must be tall enough to meet clearance height requirements. The temporary pole may need to be higher if the power company supply is on the opposite side of the street or highway.

FIGURE 2.1   Example of temporary power for new construction.

The man-made power pole needs to be braced in three directions with the rear brace in line with the service drop. The braces need to be attached 4 feet from the top of the pole, spread out at the bottom and spiked to 2 x 4 inch stakes driven 3 feet in the ground. A service cable with a weather head or gooseneck is attached to the pole. This carries power down through a meter socket to a rain tight disconnecting device. The disconnecting device must have provisions for locking to prevent unauthorized access to live terminals. The service-disconnecting device will be equipped with GFCI devices. The GFCI devices are installed on all 120 volt single-phase 15, 20, and 30 ampere receptacles for the protection of the personnel that will be using the power outlets (Figure 2.2). Grounding the temporary power panel is done by driving an 8 foot ground rod. Ground the electrical panel with #6 copper or #4 aluminum wire to the ground rod with the proper ground rod clamp.

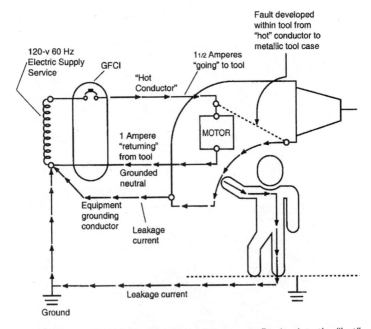

FIGURE 2.2 GFCI monitors the difference in current flowing into the "hot" and out to the grounded neutral conductors. The difference (½ ampere in this case) will flow back through any available path, such as the equipment grounding conductor, and through a person holding the tool, if the person is in contact with a grounded object.

EXTERIOR WIRING

Exterior wiring is not any more difficult to understand or install than interior wiring, but the procedure and requirements are different. One of the first concerns when working with exterior wiring is the need to protect the wiring. A lot of outdoor wiring is buried to protect it. Some wiring is run overhead. In either case, there are regulations that must be observed to conform to local code requirements.

What is outdoor wiring used for? Well, it can serve many needs. Some of the more common applications include:

- *Security lighting*
- *Walkway lighting*
- *Spot lighting*
- *Flood lighting*
- *Deck lighting*
- *Pool lighting*
- *Electrical service to outdoor buildings, such as gazebos and sheds*

- *Signage*
- *Well pumps*
- *Septic pumps*
- *Sensors*
- *Electrically-operated gates*
- *Exterior receptacles*

OVERHEAD WIRING

Overhead wiring is a common means of getting cable from one point to another when working outdoors. Running cable overhead is less expensive, usually, than burying cable. Yet, most contractors agree that an underground installation is safer.

If you are going to run exterior cable overhead, you will have to be aware of the height requirements. For example, if you are running cable over a sidewalk or a platform and the cable carries a maximum of 150 volts to ground, the cable must be at least 10 feet above the sidewalk or platform. Depending on the maximum voltage and the type of use under the cable, the height requirement increases.

If you are installing exterior cable that has a maximum of 300 volts to ground over a home, a driveway, or non-commercial traffic, the minimum height from the ground to the cable is 12 feet (Figures 3.1, 3.2).

A cable that carries voltage of 300-600 volts requires a clearance of 15 feet from the ground when it is run outside and over public streets, alleys, and roads that do not have truck traffic.

When public streets, alleys, roads that have truck traffic, and non-residential driveways must be crossed by exterior cable that carries a maximum of 600 volts, the cable must have a minimum clearance of 18 feet from the ground.

FIGURE 3.1 Typical overhead residential service connection.

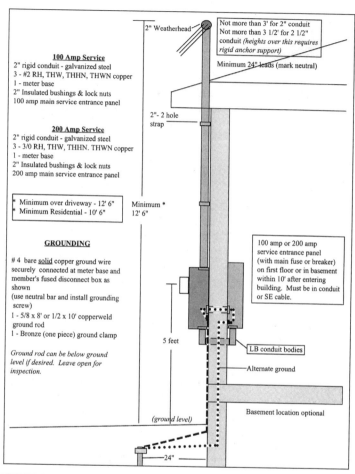

FIGURE 3.2 Overhead conduit riser service.

In addition to minimum clearances for overhead wiring, you must consider other potential risks when deciding whether to go overhead or underground. For example, will tree limbs fall on hanging cable? How often will trees have to be pruned to prevent interference with overhead cables? What are the chances that the overhead cable will be damaged in an ice or windstorm? The low initial cost of overhead wiring may not be a bargain when you factor in the long-term maintenance requirements.

GOING UNDERGROUND

Going underground with exterior wiring is a good idea. This procedure can be more expensive than an overhead service, but the cable is protected from ice, wind, and tree limbs when it is buried (Figure 3.3). There is some risk to an underground cable being dug up by mistake, but if the path of the cable is recorded and marked, this risk is nearly nonexistent.

There are three basic options in the choice of materials and procedures for installing underground cable. Direct-burial (UF) cable is the faster way to get an underground cable installed. If you elect to run the cable through a conduit, you can use a rigid, nonmetallic conduit, such as PVC pipe, or rigid and *intermediate metal conduit* (IMC).

Direct-Burial Cable

Direct-burial (UF) cable is similar to NM cable in appearance. However, UF cable is clearly marked on its sheathing. Before you install a direct-bury cable, make sure that you are working with UF cable. Sheathing for UF cable is created with plastic and provides a coating that allows UF cable to be buried without the need for conduit. However, if UF cable is run above ground, it must be protected by being installed in a conduit.

How deep does UF cable have to be buried? The depth requirement depends on the type of location where the UF cable will be installed. For example, if the burial will take place in soil where there

fastfacts

Every experienced professional should know to call and check for underground utilities prior to digging trenches. As much as this is simple common sense, there are still contractors who fail to confirm the presence of underground wires, pipes, and utilities. Many regions enforce stiff fines for damage caused to underground devices when a pre-digging request for locating any underground obstacles is not made. Before you dig, call the proper people in your area to get clearance for the digging.

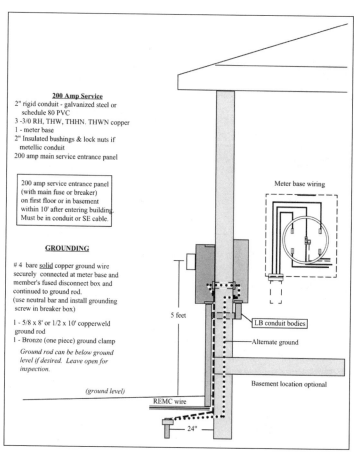

FIGURE 3.3 Residential underground service.

is only pedestrian traffic, the minimum depth for burial is 24 inches. If UF cable is buried under at least 2 inches of concrete, the minimum depth is 18 inches. A minimum depth of 24 inches is required when the cable is installed under streets, highways, roads, alleys, driveways, or parking lots. When being buried under the dwelling driveway for a one- or two-family dwelling, the minimum burial depth is 18 inches.

fastfacts

Always look for the Underwriters Laboratories (UL) rating symbol on electrical materials. Don't install products that don't carry the UL rating. You must also be sure to follow UL guidelines when installing listed products. Failure to follow this practice is a direct violation of Article 110 of the National Electrical Code.

PVC Conduit

When UF cable is installed in PVC conduit and buried, the minimum burial depths are less. For example, this type of installation in soil with only pedestrian traffic has a minimal burial depth of 18 inches. This is 6 inches shallower than what would be required without the conduit. When installed under a concrete floor that has a minimum of 2 inches of concrete, the minimum burial depth is 12 inches. Again, this is 6 inches less than what would be required without the conduit. The minimum burial depth for installations under streets, highways, roads, alleys, driveways, and parking lots is 24 inches. This is the same requirement as would be enforced without a conduit. The minimum depth for a one- or two-family dwelling driveway is 18 inches. Again, this is the same as what would be required without the conduit.

Rigid and IMC Conduit

In some cases, the use of rigid and IMC conduit reduces the burial depth even more. When buried in soil used only for pedestrian traffic, the minimum burial depth is 6 inches. When placed under a minimum of 2 inches of concrete, the minimum burial depth is 6 inches.

 Trade Tip: Rigid-metal conduit is not allowed in cinder fill unless certain conditions are met. If the cinder fill is subject to permanent moisture, the conduit cannot be used, unless it is protected on all sides by a layer of noncinder concrete.

Placing the installation under streets, highways, roads, alleys, driveways, and parking lots will still require a minimum depth of 24 inches. One- and two-family dwelling driveways require a minimal burial depth of 18 inches.

OUTSIDE COMPONENTS

The outside components of an electrical system must be, at a minimum, weatherproof. In some cases, the components must be watertight. Obviously, water and electricity don't go well together. There can be confusion about the difference between rain-tight and waterproof. Typical outdoor receptacles are weather proof. They are safe when the covers are closed, but they are not watertight and can be a risk if the covers are not closed. A true waterproof box is sealed with a waterproof gasket and can withstand a soaking rain or even temporary saturation. Choose the right box and cover for the system you are installing (Figures 3.4, 3.5).

RECEPTACLES

Exterior receptacles are required to be on a GFCI circuit. You can install GFI outlets, but these outlets can be temperamental in wet conditions. Yes, that is their job, but there can be a lot of cutoffs by the outlet that may not be needed. Some electricians will use a GFI breaker in the panel box for the circuit, but we prefer to use GFI receptacles. This way the tripped GFI can be reset at the point of usage rather than searching for the breaker in the panel box inside the house.

fastfacts

What is the minimum circuit requirement for outdoor use? Always check your local code requirements, but, normally, one 20 amp circuit is all that is required. However, the circuit must be rated as a ground fault interrupter circuit (GFI). Lighting should be installed on a separate circuit.

fastfacts

Be careful not to overload a box with too many connectors. If you do this, you may damage the insulation on wires. It's also possible that a switch or receptacle could be damaged by overstuffing a box with conductors. There are code requirements on how many conductors are allowed in a box. Follow the code regulations to avoid shorts in the electrical system. Sizing junction boxes generally corresponds to a certain volume in cubic inches per conductor size. For example, a #12 AWG conductor requires 2.25 cubic inches of volume inside a box. Therefore, a four inch square box that is 1½ inches deep with a volume of 21 cubic inches is rated for up to nine #12 wires.

Box Dimensions, inches	Maximum Number of Conductors		
	No 14	No 12	No 10
3 1/4	4	4	3
4	6	6	4
1 1/4 4 square	9	7	6
4 11/16	8	6	6

FIGURE 3.4A   Conductor chart for shallow boxes (less than 1½ inches deep).

Box Dimensions, inches	Maximum Number of Conductors			
	No 14	No 12	No 10	No 8
1 1/2 x 3 1/4 octagonal	5	5	4	0
1 1/2 x 4 octagonal	8	7	6	5
1 1/2 x 4 square	11	9	7	5
1 1/2 x 4 11/16 square	16	12	10	8
2 1/8 x 4 11/16 square	20	16	12	10
1 3/4 x 2 3/4 x 2	5	4	4	4
1 3/4 x 2 3/4 x 2 1/2	6	6	5	0
1 3/4 x 2 3/4 x 3	7	7	6	0

FIGURE 3.4B   Conductor chart for deep boxes.

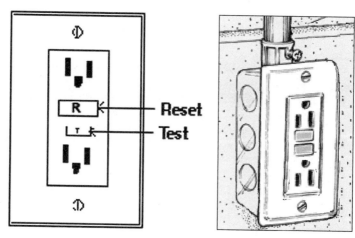

FIGURE 3.5   GFCI receptacles.

Residences are required by code to have at least one receptacle installed at the front and rear of the home. The outlets are required to be within 78 inches of the finished grade. These outlets generally have flip-up, gasket-fitted covers. However, if the outlet will be used for unattended purposes, such as running a sump pump, it must be equipped with a weatherproof box and a cover that protects the box when a plug is plugged into the outlet. These covers are available in both horizontal and vertical versions. Never use a horizontal cover in a vertical position; likewise, never use a vertical cover in a horizontal position.

Exterior outlets can be installed on walls, posts, or any secure, approved location. Receptacles that are post-mounted can be screwed to a wooden post, which should be pressure-treated to prevent rotting, or they may be attached to ½ inch conduit that is secured in concrete. The conduit must be galvanized rigid metal. A bucket of concrete that has formed around the conduit can serve as an anchor. The depth at which the bucket must be buried varies from jurisdiction to jurisdiction, so check your local code requirements for an approved depth.

SWITCHES

Exterior switches have to be installed in weatherproof boxes and are required to be fitted with weatherproof covers. There are covers available for single, double, and triple-gang boxes that are operated by toggle levers. You can also get a cover for a combination outlet-switch device.

CONDUIT

There are three types of conduit that can be used with exterior wiring. Rigid conduit is often used. Intermediate metallic conduit (IMC) is also used. Then there is PVC nonmetallic conduit. This type of conduit comes in Schedule 40 PVC and Schedule 80 PVC. Not all locations allow its use, but most do. Any connectors and fittings used will have to conform to code requirements for the type of conduit being installed.

EXITING A HOME WALL

Exiting a home wall with conduit will require a special L-shaped connector that is known as an LB conduit body. This is a 90 degree fitting that is meant to house UF cable. This device contains the junction between interior cable and exterior cable. The fitting allows a transition and delivers the UF cable to an underground location. An LB conduit body is fitted with a gasket that seals the cable connection against weather conditions. An LB conduit body has a removable cover on the back (Figure 3.6). In special applications, an LB opening may not be accessible. There is a version available with the opening on its right side. This is called an LR (Figure 3.7). A version available with the opening on its left side is called an LL (Figure 3.8).

EXTERIOR LIGHTS

Exterior lights run the gamut of styles, sizes, and shapes. Code requirements call for exterior lighting at all outdoor entrances to homes. These lights must be controlled by switches. The lights are required for all entries and exits accessible from ground level, at attached

FIGURE 3.6   LB conduit body.

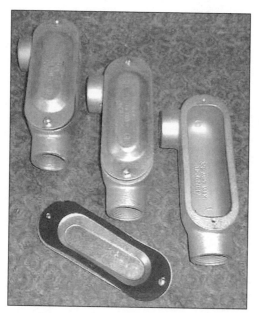

FIGURE 3.7   LL conduit body.

FIGURE 3.8 LR conduit body.

garages, and at detached garages that are equipped with electrical power. There are, of course, many other situations where exterior lighting might be wanted. Here are some examples of such uses:

- *Pool lighting*
- *Decorative lighting*
- *Pond lighting*
- *Deck lighting*
- *Walkway lighting*
- *Security lighting*
- *Spotlighting*
- *Floodlighting*
- *Garden lighting*
- *Accent lighting*
- *Driveway lighting*
- *Play area lighting*
- *Party lighting*

The two most common types of lights used outside are reflector-type (R-type) lights and parabolic aluminized reflector (PAR) lights (Figures 3.9a, b). Both of these types of lamps have a long life and a reflective interior surface. They also resist weathering. Of the two types, PAR lamps are better suited for outdoor use than some kinds of R-type lamps. If you plan to use R-type lamps, check to see that they are approved for the intended use. Outdoor lamps of these types are used in outdoor lamp sockets that have gaskets to make the installation weatherproof.

Many types of outdoor lighting run on 120 volt systems, but there are also a number of fixtures that operate on low-voltage wiring. If low-voltage wiring is used, a transformer that will convert 120 volt current to 12 volts is needed. Walkway lighting is a good example of where you might use low-voltage wiring.

AUTOMATED LIGHTS

Automated lights could be controlled by sensors or by timers. Motion-sensor floodlights and spotlights are good choices to deter burglars. The same lights are convenient when you pull into your driveway after dark. If outdoor lighting is connected to a home-automation system, this part of the equation is already solved. If not, you have to decide between sensors and timers.

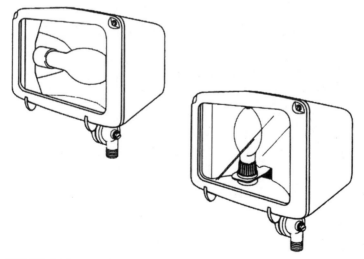

FIGURE 3.09a Metal halide outdoor floodlights utilizing parabolic aluminized reflectors. *(Courtesy Lithonia Lighting)*

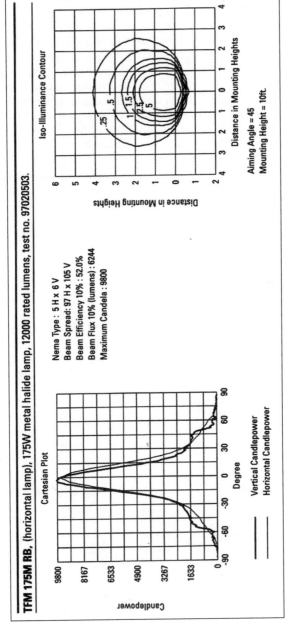

FIGURE 3.09b Photometric data for metal halide outdoor floodlights with parabolic aluminized reflectors. *(Courtesy Lithonia Lighting)*

Motion-detector lights are designed to come on when motion is detected. Most sensors are a built-in part of the lamp holder. A photo-cell shuts the light down during daylight hours. Most motion-detector lights can be programmed to remain on for a set amount of time. A typical manual switch can also be used to operate a light. Other sensors can turn lights on when light levels outside fall to a certain level and turn the lights off when there is sufficient natural lighting. This type of system is often used for driveway and walkway lights.

Traditional timer switches can be used to control outdoor lights. General opinion states that timers should not be used on lamps with a wattage rating in excess of their full load rating. Timers can also be used to control multi-circuit lighting contactors. Some timers are not designed to prevent damage from overloading. The best choice for pre-programmed lighting is an automated home system.

GETTING OUT

Getting out of a house to supply power to exterior electrical devices is usually not difficult. The circuit originates at the panel box. Ideally, the cable will exit through the bandboard of the home. An LB fitting will be used at the exit point, and the center of the exit hole is required to be at least 12 inches above finished grade and no more than 78 inches above the ground.

Interior wiring is run to a junction box within the walls of the dwelling. This is where the UF cable is joined to the interior wiring. A conduit nipple is used like a sleeve through the wall of the home. The conduit nipple is attached to an LB fitting. Rigid conduit goes down into the trench and is fitted with a rigid conduit sweep bend. Then the UF cable is run, and expansion loop is left where the cable enters the sweep bend. This is about all there is to getting wire to work with outdoors.

Running wire for outdoor fixtures is not much different than general wiring. There are, of course, special considerations such as weatherproofing. Overall, the work is not particularly difficult.

4

NEW CONSTRUCTION

Some electricians specialize in new construction. The work is very different from remodeling and service work. Some people prefer one type of work to another. Most electricians agree that new construction work is much easier, in most cases, than working with existing systems. Other electricians find the routine of construction work boring. What do you prefer?

I have done remodeling, service work, and new construction. There are parts of each type of work that I enjoy. Remodeling work is challenging and often pays very well. Service work allows you to move from job to job, so you don't get tired of going to the same job day after day. New construction is clean and usually simple. They all have their pros and cons.

Every type of work has a certain rhythm to it. If you are doing commercial work, you will have detailed wiring diagrams and production schedules to follow. Residential work, however, usually puts a lot more of the responsibility of planning on the electrician. In either type of work, you have to be organized and have a plan. This can make all the difference in how well a job goes.

If you go to a house that has been framed and string wire all through it before the HVAC contractor gets on the job, you are probably going to have a big problem. Your wires can be routed in many places and ways. Ductwork is not as flexible and requires certain areas for installation. If you have filled a duct chase with wires, you will be moving them. This same type of problem could arise if you go in before the plumbing contractor. Therefore, you have to coordinate your work well with other contractors when doing construction work.

As an electrician, you are likely to be one of the first contractors on a construction site. The site contractors will get there before you to clear trees, make a driveway, and dig a foundation. A well-drilling contractor may beat you to the job. But, there is a good chance that your services will be requested early on to install a temporary power pole. Some jobs are run with generators, but temporary power poles are normally installed for workers to power their tools.

MEETINGS

Before you start a construction job, you should have meetings with the general contractor and a representative of the local utility company. You will need to establish basic information on where the service panel will be located, whether the incoming service will be overhead or underground, and where a temporary pole will be placed. Blueprints may show you where the service panel is wanted, but it still pays to do an onsite meeting with the builder and a utility representative to ensure that the planned location is viable.

Once you know if the power supply will be installed overhead or underground, you will be able to plan your materials for an underground service coming from the pole or an overhead service with or without a mast. Chances are good that you asked about this before you put a bid on the job, but confirm the information before going ahead with bid specs. After you have all of your facts in order, you can begin your preliminary work.

PERMITS

Before you begin work, you will need to obtain an electrical permit from your local code enforcement office. This normally requires the submission of two sets of drawings and specifications. Once the code office approves the documents, you will be given an approved set of documents that will bear a stamp from the code office. Most jurisdictions require you to keep a set of approved plans and specifications on the job. You will also be issued a permit.

fastfacts

Many jurisdictions require permits to be posted so that they can be seen from the road that runs past the property.

TEMPORARY SERVICE

Setting up temporary service is very helpful for people working with power tools. Most electricians make up temporary service poles off-site and bring them to a jobsite when they are needed. Some electricians build the power poles on the jobsite. It is recommended that the service be rated for at least 50 amps at 240 volts. The temporary board should have at least 220 amp GFI protected receptacles. You also should plan on a couple of 240-volt receptacles for special equipment such as mortar mixers, block saws, welders and large air compressors. Either way, the pole will need to be inspected and approved before it can be used.

ORDERING MATERIALS

Ordering materials for a job is an important part of making the job successful. If you go to a job without the right materials, you will lose a lot of time running back and forth to a supply house. You probably did a takeoff from the blueprints when you bid the job, but do a walk-through of the framed building before ordering your materials. It is not uncommon for things to change from what was anticipated on blueprints. This is especially true of residential construction.

You may be thinking that a walk-through to check on materials is a waste of time. It is not, and you can make the walk-through even more productive. You are going to have to mark installation locations at some point; you might as well do that while you are double-checking the material needs. Then, when you arrive to rough-in a job, the marking is already done and you and your helper can get right to work. By marking locations at this point, your take-off will be dead on, although you should still order a little more material than you think you will need. Leaving the job to run materials is a big waste of time that hurts a lot of contractors. When you buy a material package for a job, you will get better pricing on individual items. If you run short of a particular item, it will generally cost more than it did with the lump material package; therefore, it is wise to order extra material. Anything left over will eventually get used.

SCHEDULE

Get a schedule from the general contractor for all of your work. Are you going to need to install conduit in the ground before a basement floor is poured? Will you first work on the job when the home is ready for rough-in? When will you set up the service panel and meter socket? If you are renting the temporary power pole to the contractor, the contractor will probably want you to set the panel and meter socket as soon as possible. When will you do exterior wiring? How long will it be before you trim out the job and hang fixtures?

A good builder can give you a schedule, but it will probably change. Still, do your best to know and maintain a viable schedule. It is best to stay one step ahead of the job if possible. The electrician is usually the last person to finish the job because wall covers and surface fixtures have to be put on after all other work is finished. This leaves you to blame if the job is behind schedule, even though you couldn't finish until the other trades were done their work. Being able to go in and finish quickly without any major tasks left to complete will leave you with far fewer headaches.

BASEMENT AND SLAB WIRING

Basement and slab wiring that will be covered with concrete has to go in early. If you are going to be running conduit for an under-floor installation, you need to get the work done and inspected prior to the floor being poured. You may be competing for space with the plumbing contractor and possibly the heating contractor. Talk to the

Don't Do This! Don't use type AC armored cable for wet locations. If you are going to use armored cable in a wet location, use type MC cable.

Don't Do This! Don't do a direct-burial installation with type AC armored cable. Type MC cable that
is labeled for direct burial can be used as substitute.

builder to coordinate your pathways and schedule for installing the
underground work.

EXTERIOR WIRING

Exterior wiring can include anything from wiring for a submersible
well pump to walkway lighting. You will want to get all trench work
done prior to final landscaping. This is another situation where you
must work with the builder to establish a schedule. When you install
underground wire, mark your path clearly to avoid having any other
workers damage your work. This is done most effectively by laying
red caution ribbon over the buried electrical lines, usually 6 to 12
inches below finished grade.

When planning your trench work, take into consideration the
plumbing contractor's need to install a sewer and water service. Talk
to the builder or plumbing contractor to avoid conflicts. You will also
have to consider the need for drain tiles, irrigation systems, and so
forth. Be careful whenever you drive a ground rod near the building.
Make sure you are clear of any underground piping or drainage tiles.

fastfacts

*When adding to an existing system or installing a new system,
you should create a load sheet. This is a worksheet that details the
types of electrical devices that will be used on the entire electrical
system. By dividing watts by voltage you will arrive at the number
of amps required for various circuits. Working from a solid load
sheet will reduce the risk of installing inadequate wiring systems.
This should be calculated before the work begins. Add 15 to 25
percent for future expansion.*

Don't Do This! Don't try to save your pennies while losing dollars. Digging a trench with a shovel will usually work, but don't dig it yourself. Calculate the cost of your time as an electrician against the cost of a laborer and see which way the job shows the most profit. Consider using a trencher. You can rent one. Not only will the trench be narrower and less noticeable, it may be less expensive when the job is done.

ROUGHING-IN

Roughing-in a house is rarely challenging. However, you do have to pay attention to what you are doing. There are also times when circumstances take the routine out of the work and make you think. After the HVAC and plumbing contractors complete their rough-ins, it's your turn. If you have planned your work well, the rough-in for a typical house goes quickly.

Most electricians put their helpers to work with drills while the electrician does more layout work or begins the rough-in wiring. Make sure to keep holes near the center of wood framing members when possible. Don't cut and notch joists, except as a last resort. Neatness counts on a job, so route your wiring carefully. Holes drilled through studs should be all at the same level. This will make it a snap to pull the cables through. It also makes the job look super professional to the general contractor and the owner.

Hanging boxes and pulling wire is what most of a rough-in is all about. Seasoned electricians find this type of work almost mechanical. They have done the work so many times that they don't really have to think about it. This is one reason why some electricians don't like new work.

fastfacts

How much wire should be left hanging at a box when roughing in an electrical system? General practice calls for 6 to 8 inches of wire to be left at the box. An old electrician always used to say, "If you don't have enough to cut off, you don't have enough."

Trade Tip: When roughing-in boxes in uncovered walls, you must allow for the depth of the wall covering. Drywall is the most common type of wall covering. You can carry a scrap of drywall with you to use as a guide when setting boxes. Some boxes have gauge lines that you can use as a reference point when setting the boxes. In any case, remember to take the rough-in measurements seriously and double check the less-experienced workers.

Housings for bathroom ventilation fans and recessed wall heaters need to be installed during a rough-in. Venting material for the bath fans should be run during the rough-in phase. Venting provisions for range hoods should also be installed during the rough-in. Essentially, any fixtures or devices that require venting or recessed housings have to be roughed in before a framing inspection is done.

Once all electrical components such as outlet circuits, lighting circuits, water heaters, and HVAC units are roughed in, you are ready for an inspection. You must have the rough-in work inspected before it is concealed. Since the electrical contractor is usually the last mechanical contractor on a job during the rough-in phase, you may get some pressure from the builder to get your inspection so that insulation can be installed. Call for an inspection promptly after completing and checking your rough-in.

fastfacts

When roughing-in a box for a ceiling fan you must determine the weight load required of the box and support arms. If you don't know how heavy the ceiling fan will be, find out. Ceiling fans vibrate and move when they are running, so the box and support must be rated to handle the fixture being used. Always try to support your fan bracket with multipurpose screws to a heavy framing member such as a floor stringer, beam, rafter or strapping.

fastfacts

Nonmetallic cable should be stapled to a solid support at minimum intervals of about 54 inches. The cable should also be stapled within 6 to 12 inches of each electrical box. I recommend 6 inches. Remember to always check your local code requirements for specific support intervals. Also, keep wire well out of the range of screws and nails. National code requires 1½ inches from the surface of framing members, I recommend 2 inches or more if possible.

Trade Tip: When working with metal studs, you should install bushings prior to pulling wire through the studs. Since the studs are metal, rough edges around a hole could cut and compromise the insulation of a wire.

TRIMMING OUT

Trimming out a job takes time, but it is not usually a complicated process. Installing devices like switches and outlets is fairly simple, but the process takes time. Installing breakers in a panel box can be dangerous, but it is not difficult for a licensed electrician. Wiring equipment and hanging fixtures take time and occasionally create problems. Sometimes fixtures are received damaged. This can slow down a job. It's a good idea to inspect fixtures before putting a trim crew on a job, but this is not always practical.

Exterior wiring for lights is no big deal. Installing landscape lighting, walkway lighting, driveway lighting, and other forms of exterior trim-outs are one of the last steps in completing a job.

KEEPING ORDER

Working in an orderly fashion is a key element of successful and profitable jobs. Plan your work and work your plan. You can bet that some

fastfacts

Plastic electrical boxes are most often used for protected indoor wiring. They are used with NM cable and are not suited for heavy fixtures and fans. They are permitted for support of wall and ceiling fixtures not exceeding six pounds in weight.

things will go wrong. This will require you to adjust your schedule and may affect other jobs. Good management is almost as important as safe wiring. Don't rush a job by cutting corners, but set goals for yourself and reach them to ensure a long and fruitful career.

5

INSTALLATION TIPS, TRICKS, AND TECHNIQUES

The most experienced electricians can also use new tips, tricks, and techniques to add to their professional ability. Routine wiring is taught in classrooms and on the job, but where to do you get the inside track? If you are lucky enough to work with a pro who will train you, the knowledge will come from the expert. On-the-job training and face-to-face communication is the ideal way to expand your skills, but books are also a great source of valuable information that is both affordable and accessible.

Since it is assumed that you are already a licensed electrician, we won't bore you with the basics of circuits and wiring applications. Instead, we will give you tips that you can learn from or refresh your memory with. You probably have heard the old saying that goes like this: "He's forgotten more than you will ever know." Well, there is some truth to that saying. People do forget some of their trade skills if they are not used frequently. It never hurts to brush up on the basic fundamentals of your trade. So, let's get started.

SIZING CIRCUITS

Sizing circuits is necessary whether you are remodeling a building or working with new construction. A 15 amp fuse or circuit breaker can be wired with 14 gauge wire. If a 20 amp fuse or circuit breaker is used on a circuit, the cable should be 12 gauge. A 30 amp fuse or circuit breaker should be used with 10 gauge wire. When you are planning a job, you must make sure that the circuits you will install are sufficient for the load that will be put on them. Don't fail to size circuits before you run your cables. As a reminder, you can determine amperage by taking a device's wattage rating and dividing it by the circuit voltage. The total amperage load for a circuit should not exceed 80 percent of the circuit breaker or fuse rating (Figures 5.1 through 5.4).

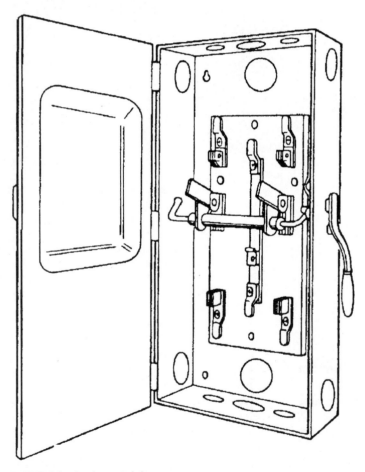

FIGURE 5.1   Service switch box.

FIGURE 5.2 Circuit breakers.

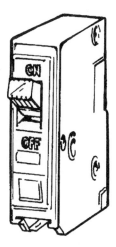

FIGURE 5.3 Single pole circuit breaker.

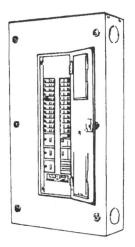

FIGURE 5.4 Panel board containing several circuit breakers.

Average Appliance Wattage Ratings

✔	Attic fan	400 watts
✔	Blender	400-1,000 watts
✔	Broiler	1,400-1,500 watts
✔	Can opener	150 watts
✔	Central air conditioner	2,500-6,000 watts
✔	Clock	2-3 watts
✔	Clothes dryer	4,000-5,600 watts
✔	Clothes washer	500-1,000 watts
✔	Computer and monitor	565 watts
✔	Coffee maker	600-1,500 watts
✔	Crock pot	100 watts
✔	Deep fat fryer	1,200-1,600 watts
✔	Dehumidifier	500 watts
✔	Dishwasher	1,000-1,500 watts
✔	Electric blanket	150-500 watts
✔	Electric water heater	2,000-5,500 watts
✔	Exhaust fan	75-200 watts
✔	Floor polisher	300 watts
✔	Food freezer	300-600 watts
✔	Food mixer	150-250 watts
✔	Frost-free refrigerator	400-600 watts
✔	Electric frying pan	1,000-1,200 watts
✔	Gas furnace	800 watts
✔	Garbage disposer	500-900 watts
✔	Hair dryer	400-1,500 watts
✔	Hot plate	600-1,000 watts
✔	Iron	600-1,200 watts
✔	Laser printer	1,000 watts
✔	Microwave oven	1,000-1,500 watts
✔	Oil furnace	600-1,200 watts
✔	Portable heater	1,000-1,500 watts
✔	Radio	40-150 watts

fastfacts

You can use a simple formula to determine the number of amps to be placed on a circuit. Refer to rating data on the devices that will be used on the circuit. For example, you might find that a refrigerator is rated at 500 watts. A toaster might use 1,050 watts. Add up the wattage for all devices to be placed on a circuit. Divide the total wattage by 120 (this represents the number of volts on the circuit). The answer to the equation is the number of amps that will load the circuit. Watts are now referred to as volt-amperes as they are a product of volts and amperes.

✔ Range	4,000-8,000 watts
✔ Range oven	3,500-5,000 watts
✔ Refrigerator	150-300 watts
✔ Roaster	1,200-1,650 watts
✔ Room air conditioner	800-2,500 watts
✔ Sewing machine	60-90 watts
✔ Stereo	50-140 watts
✔ Television	50-450 watts
✔ Toaster oven	500-1,450 watts
✔ Waffle iron	600-1,200 watts

CHOOSING THE RIGHT BOXES

Choosing the right boxes for electrical devices can be a bit more complicated than you might think. There are a lot of options available, and some types of applications require specific types of boxes. With this in mind, let's do a quick rundown on the boxes that you might be dealing with (Figures 5.5 through 5.11).

FIGURE 5.5 Single gang plastic electrical box.

FIGURE 5.6 Two gang plastic electrical box.

Size of Conduit, inches	Maximum Distance Between Rigid Metal Conduit Supports, feet
1/2	10
3/4	10
1	12
1 1/4	14
1 1/2	14
2	16
2 1/2	16
3	20

FIGURE 5.7 Support for rigid metal conduit runs.

Size of Conduit, inches	Conductors Without Load Sheath, inches	Conductors With Load Sheath, inches
1/2	4	6
3/4	5	8
1	6	11
1 1/4	8	14
1 1/2	10	16
2	12	21
2 1/2	15	25
3	18	31
3 1/2	21	36
4	24	40
5	30	50
6	36	61

FIGURE 5.8 Radius of conduit bends.

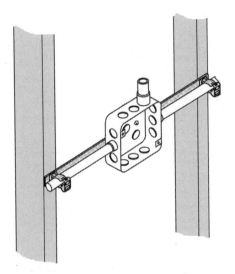

FIGURE 5.9 Box and conduit mounting strap.

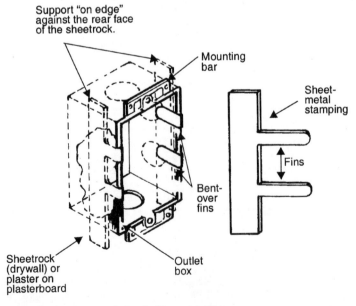

FIGURE 5.10 Patented electrical box supports.

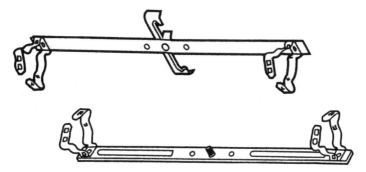

FIGURE 5.11   Bar hangers.

Standard Boxes

Are there really any standard boxes in the electrical trade? Not exactly, but plastic boxes that have depth gauges on them and have nails attached in a nailing channel are, by far, the most popular type of switch or outlet box used in residential applications. Most of these boxes have a depth of 3½ inches. The convenient boxes come in single gang, double gang, and four gang styles. Plastic boxes that are double gang or four gang require internal cable clamps.

Plastic Retrofit Boxes

Plastic retrofit boxes are often used when a new switch or outlet is wanted and the box must fit inside of a finished wall. Internal cable clamps are used with these boxes. There are also plastic retrofit boxes designed for use with light fixtures. These boxes can be installed in existing walls and ceilings.

Plastic Boxes For Lights

Plastic boxes for lights are also very popular. These are available with metal brackets that will span the distance between rafters or joists to allow for centering of light fixtures.

Metal Boxes For Lights

Metal boxes for lights come in heavy-duty versions. These are useful when installing ceiling fans and heavy light fixtures. Like their plastic counterparts, the metal boxes can slide along the brace bar to center a fixture. Standard metal boxes designed to support ceiling fixtures and fans that are generally limited to 50 pounds suspended load. Loads in excess of 50 pounds must be supported from the building structure (Figure 5.12).

Metal Outlet and Switch Boxes

Metal outlet and switch boxes are used with exposed indoor wiring. They are commonly used with metal conduit. Since the boxes are metal, they must be grounded (Figures 5.13 and 5.14).

FIGURE 5.12   Metal octagonal ceiling box.

fastfacts

Metal electrical boxes are used for exposed indoor wiring, with metal conduit, with protected indoor wiring, and with NM cable. Cast-aluminum electrical boxes are for outdoor wiring and can be used with metal and non-metallic conduit. PVC plastic boxes are good for outdoor wiring, exposed indoor wiring, and can be used with PVC plastic conduit.

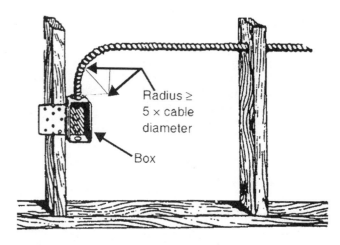

FIGURE 5.13 Armored cable bend with connection to outlet box.

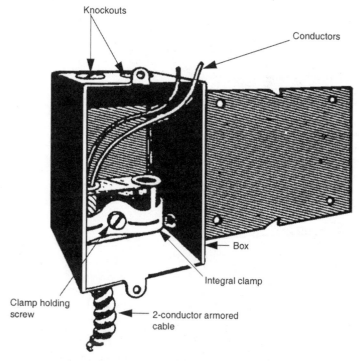

FIGURE 5.14 Cable connections to box with internal clamps.

Cast-Aluminum Boxes

Cast-aluminum boxes are required for outdoor fixtures connected with metal conduit. They have sealed seams and threaded openings to keep moisture out. Many types of waterproof coverplates are available for these boxes. You can get coverplates for duplex outlets, GFI outlets, and switches. Apply a thin coating of an antioxidant compound to all threaded fittings when working with cast aluminum boxes.

BOX PLACEMENT

Box placement for outlets and switches can vary, but there are general rules-of-thumb that are used to determine the locations of various types of boxes. Switch boxes are normally installed 48 to 50 inches above the floor. When wiring to handicapped standards, a switch must not be more than 48 inches above the floor. Outlet boxes are normally set 12 to 16 inches above a finished floor. Receptacle boxes should be set 18 inches above a floor when wiring a facility to handicapped standards. Outlet boxes in laundry rooms are normally installed 42 inches above the floor. Outlet boxes for wall brackets and sconces should be placed 5½ to 6½ feet above the floor.

When attaching boxes in new construction, hold the front edge of the box out to allow for wall coverings. Boxes should not be recessed more than ¼ inch from the finished wall surface. If combustible materials are present, an electrical box should be mounted flush with the material. All boxes must be securely attached. Additionally, the wall material should not be cut out more than ⅛ inch from the perimeter of a recessed box.

fastfacts

If you are planning to use PVC plastic conduit, you may use plastic or metal boxes. A green ground wire should always be run with the conduit. Metal boxes must be bonded to the green ground wire.

SURFACE WIRING

When remodeling or working in existing structures, you may need to install surface wiring. This is done with a surface-mount box and raceway to conceal the wiring. If plastic raceway is used, a separate ground wire must be run with the cable. Metal raceway that is connected to a properly wired and grounded electrical box is self-grounding and eliminates the need for an individual ground wire being run with the cable. A smart electrician always runs a ground wire with a circuit. You never know when a raceway fitting could break or be pulled loose. This would cut off the grounding path of the raceway system. With the installation of a separate grounding conductor, you can always be assured of a reliable ground path in the event of a ground fault (Figures 5.15 through 5.17).

NAIL PLATES

Nail plates are required when holes that contain cables are within 1¼ inches of the nailing surface of a framing member. The most efficient type of nail plates has sharp drive points as an integral part of the plate. This allows you to simply hold the plate in place and drive its drive points into a framing member (Figure 5.18).

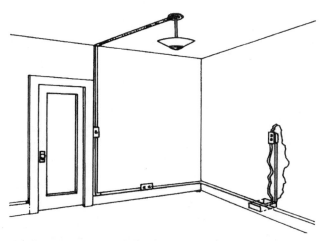

FIGURE 5.15 Surface metal raceways.

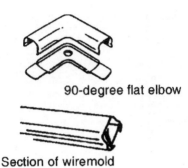

90-degree flat elbow

Section of wiremold

FIGURE 5.16   Metal raceway connectors.

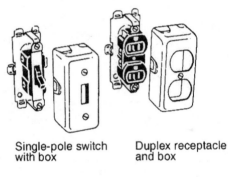

Single-pole switch
with box

Duplex receptacle
and box

FIGURE 5.17   Surface raceway outlet and switch.

fastfacts

*Receptacles above countertops must be no more than 4 feet apart.
Any section of countertop that is more than 12 inches wide is required to have a receptacle installed above it. When counters are
separated by appliances, the counter areas are considered to be individual counters. All receptacles in bathrooms are required to be
GFI protected. All receptacles in kitchens within six feet of the sink
are required to be GFI protected.*

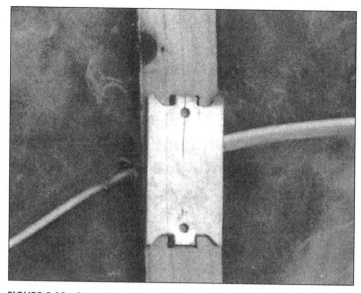

FIGURE 5.18   Protective plate used to prevent nails and screws from hitting vulnerable materials.

ATTICS

When running cables in attics, you must protect any wires that are within 6 inches of any attic-access openings. Most electricians run a wide board along the attic joists and then secure their wires to the board. This keeps the wiring neat and in one place.

CRAWLSPACE INSTALLATIONS

Crawlspace installations can be a real pain, due to limited workspace. It is common to install a wiring board to the bottom of floor joists in crawlspaces. This provides a good surface for stapling cables. As in an attic installation, this approach keeps wiring neat and in one location. Large cables can be stapled directly to the floor joists. No cables should be left sagging. Another reliable method for running cables in crawl spaces is to drill a series of holes through the floor stringers. Cables are then threaded through the holes. The holes must be in the

fastfacts

When sizing an attic fan, multiply the square footage of the attic by 0.7. If the roof is dark in color, add 15 percent to the number. This will tell you how many cubic feet per minute (CFM) the fan should pull.

center of the stringer width. Always drill a few more holes than you think are necessary. This way, if you have to install more wires, you won't need to break out the drill again.

WORKING WIRE THROUGH EXISTING SPACE

Working wire through existing space is much more difficult than wiring open walls in new construction. Remodeling work can test the skill of any worker in any trade. Adding lights, switches, receptacles, and other electrical devices in an existing building can require some ingenuity. There are, however, some tricks of the trade that make the work easier.

If you are working on a job that has an accessible attic and basement or crawlspace, you should be able to fish wire through walls fairly easily. However, you might encounter fire blocking between studs that will block your path. If there is fire blocking, you have a problem. Let's start by assuming that there is no blocking in the walls to stop your wires. When this is the case, you have to get below finished living space and find the wall plate. In a crawlspace or unfinished basement, you might be able to see nails in a row that indicate the presence of a sole plate. If you see this type of evidence, measure from a known point, such as an outside wall, to the anticipated wall location. Then you can go upstairs and measure to see if the wall is where you think it is.

If you are having trouble locating a sole plate from below, there is another option. Remove the shoe mold or baseboard along the wall that you want to chase your wire through. Take a very small drill bit and drill a hole through the floor where the edge of the plate is. Stick a small wire, or other small thin item, through the hole to make its location easier to find when you go into the unfinished space below. Once you locate the hole, you can measure from it to find the center of the sole plate. Now all you have to do is drill up through the plate. The small hole you made to identify the wall location will be hidden when you replace the shoe mold or baseboard. Sometimes removing baseboard can be more bother than it is worth. A useful method for locating a wall without removing the baseboard is to drive a nail through the floor, next to the location where you want to drill a hole. When the nail is driven nearly full length, cut the head off with your linesman pliers. After you locate it in the crawlspace, it can be pulled out from below. It will give you a reliable reference point to measure from.

If you are simply installing a receptacle, your difficult work is basically done. You can cut a hole in the drywall with a keyhole saw to mount your box, and the wire should be within your grasp. If you are wiring a switch, you may have to use a fish tape to work the wire up to the box location.

When you need to pass through a wall into an attic, you will have to get into the attic and locate the top wall plate. By measuring in the finished space from a known point, like an outside wall, you should be able to locate a wall in an attic with ease. After finding the wall, you will need to do some more measuring to make sure that the hole you drill in the top plate will put you into the same stud chase that the lower hole is in. Remember, it is better to measure twice and drill once than to measure once and drill twice. An unwanted hole drilled accidentally up through a floor or down through a ceiling will not please your customer. Once you have both holes in the same stud bay, you can use a fish tape to pull wire through the chase.

Trade Tip: When estimating the amount of wire needed for a job, you will have to calculate the running distance of the wire. Add one foot of wire for each junction that you will encounter. Then add 20 percent to the total amount of wire you anticipate needing. For example, if you were going from an existing outlet to two new outlets over a distance of 20 feet, you would have 20 feet of running wire and 5 feet of wire for junctions. This would be 25 feet of wire. Then you would add 20 percent to this figure, which would be 5 feet, resulting in a total need of 30 feet of wire.

Trade Tip: When running wire in finished space, you might be able to gain a wiring path by removing the baseboard trim. Use a utility knife to separate the top edge of the trim from the wall. Once this is done, a pry bar is used to remove the baseboard. Now you can cut out a section of drywall that was behind the trim. This will give you a horizontal path to run wires through. After drilling studs, installing wiring, and getting an approved inspection, you can replace the baseboard to hide your work. This procedure can reduce the damage done to a finished living space.

Horizontal Runs

You know how to make vertical runs without damaging existing walls, but do you know how to make horizontal runs without damaging finished walls or using raceway? It can be a fairly simple process.

Removing the baseboard trim on a wall can allow you to run wire horizontally in finished space with no damage to the existing walls. Before removing the baseboard, use a pencil to draw a thin, light line along the top edge of the trim. Then use a utility knife to make the trim come off more easily. Insert the blade behind the trim and pull it along the trim to separate the trim from the drywall. Next, remove the baseboard. Measure down, towards the floor about ½ of an inch and snap a chalkline so that you have a level line running about ½ inch below the line you drew along the top of the trim.

After snapping a line, you can cut out the drywall below the chalkline. You can either cut a section out, leaving drywall near the floor and above the chalkline, or you can cut out the complete lower section. If you do this, save the sections so that you can replace them after your wiring work is done and inspected.

Once you have the drywall cut, you will expose the studs. This will allow you to either notch or drill them to install cable. You will have to install nailplates on each stud to protect the wiring if the cable is less than 1¼ inches from the face of the studs. When you are done, you can put the baseboard back up. Some touch up work on nail heads and paint will likely be required, but the damage done to finished living space by using this method is minimal.

CUTTING HOLES

Cutting holes is an inevitable part of remodeling work. Sooner or later, you are going to find yourself in a situation where your only choice will be to cut access holes in walls or ceilings. It helps to keep the holes to a minimum. Try to make your cuts within a stud or joist bay. This will allow the person who repairs the holes to make nice, even cuts back to the framing members to ensure a good fit and seam when the replacement drywall is installed.

Try to avoid cutting holes when you can. If you are making a horizontal run, you might be able to use a flex-bit in a drill to cut back on the number of access holes that will be needed. A fish tape is often needed for pulling wires from hole to hole.

MIDDLE-RUN RECEPTACLES

Middle-run receptacles are wired with a cable coming from a previous outlet to the outlet box and another cable leaving the box to go to another receptacle. It is common for the two cables to be connected directly to the screw connectors on the outlet. This is safe and legal, but there is a better way to do the job. Wire the hot wires from the two cables with a short jumper wire and secure them with a wirenut. Run the hot wires to the connection point on the outlet. Do the same with the neutral wires. When wired this way, the only power running through the outlet is the power going to serve the device plugged into the outlet. If the outlet is removed for any reason, the circuit will continue to operate. If the job is wired with the outlet being a basic coupling to the two cables and the outlet is removed, the circuit is not going to work. This procedure takes more time and many electricians don't invest their time in wiring outlets to provide for the potential removal of an outlet. You don't have to, but this is a good way of wiring the devices.

 Don't Do This! Don't use soap, detergent, oil, or grease as a lubricant when pulling wire. There are special pulling lubricants available—use them. The use of unapproved lubricants can result in damaged insulation on wires.

Trade Tip: When cutting raceway, you should cut backing and capping before separating the material. However, if the backing of the raceway passes under the fittings and is not cut, you have to avoid cutting the backing and capping first.

PUSH-IN TERMINALS

Many outlets are equipped with both screw type connectors and push-in terminals. The push-in holes will accept 14 gauge copper wire. It's tempting to take advantage of these timesaving outlets, but you might be setting yourself up for trouble. There are many stories about how push-in connections simply don't hold up well. Use your own judgment, but saving a few minutes on the job could cost you hours of work later. I strongly recommend that you secure wires under screw connectors and never use push-in terminals.

SPLIT-CIRCUIT OUTLETS

Split-circuit outlets have a metal tab that connects the two screw connections together. If you remove the tab that connects the two brass contact points, you can create a split circuit. A split-circuit duplex outlet allows two appliances to be plugged into the receptacle, even when the two appliances would overload a standard outlet. This is the reason for creating the split circuit. Since the circuit is split and the outlet is serving two circuits, a higher amp load can be placed on the receptacle. The two circuits supplying the split-wired receptacle must be connected to a two pole circuit breaker. Neutral conductors in a multi-outlet system must be pigtailed. Do not try to wire all of the neutral conductors on the receptacle terminal screws.

RECESSED LIGHT HOUSINGS

Recessed light housings come in two forms. If you will be installing this type of unit in an attic, or some other place where the housing may come into contact with insulation, you have to make sure that you have insulated-ceilings (IC) fixture housing. A non-insulated-ceiling

(NIC) housing must have a minimum of 3 inches of clear space between it and insulation. Don't forget to confirm that you have the proper housing for your application. It is a good idea to always use IC recessed housings because eventually insulation could be blown in around your housing.

DISHWASHERS

Dishwashers can be hardwired to an electrical system. They require their own circuit. Don't forget to leave excess cable under the appliance. It's not unusual to find a dishwasher where the electrical wiring is run fairly tightly to the appliance's connection box. This is safe enough, but it's very inconvenient if the dishwasher has to be pulled out of its under-cabinet location. Leave a few feet of slack cable under the appliance. The slack wiring allows the dishwasher to be pulled out if there is a need to do so. This is much more convenient than having to disconnect the wiring to move the dishwasher.

CEILING FANS

The electrical boxes installed for ceiling fans must be secured well enough to support the weight of the fan. When a fan weighs less than 35 pounds, you can usually get by with only the fixture box holding the fan, but check your local code requirements to be sure. Heavier fans must be supported independently of an electrical box. The most logical means of support is a support bar that spans between rafters or joists to accommodate the weight of the fan.

SIZING WHOLE-HOUSE FANS

Sizing whole-house fans is not difficult. The first step is to find the total square footage of the house that the fan will be servicing. Once you know this, multiply the total square footage by the ceiling height to find the total cubic feet of the structure. In a normal home, the ceiling height can be figured as 8 feet. If you were dealing with 3,000 square feet of floor space, you would have 24,000 cubic feet to ventilate.

When you know the number of cubic feet to be ventilated, divide that number by the number of air changes you want per minute. Let's say that you want a full air change once every 20 minutes. In this case, you would divide 24,000 by 20 and arrive at an answer of 1,200 cubic feet per minute (CFM). Now all you have to do is check a fan-sizing chart to see what size fan you need. This particular job would require a 48 inch fan.

Fan-Sizing Data

24 inch fan	3,500 to 5,500 CFM
30 inch fan	4,500 to 8,500 CFM
36 inch fan	8,000 to 12,000 CFM
42 inch fan	10,000 to 15,000 CFM
48 inch fan	12,000 to 20,000 CFM

METAL CONDUIT

Metal conduit can slow down a job, but sometimes it is needed. When you are measuring conduit for a cut, remember that it will slide into a fitting by about one inch. Cut conduit with a hacksaw. You might be tempted to cut it with a tubing cutter, but this practice is generally frowned upon. Why? Because a tubing cutter tends to bevel the cut and can create sharp edges on the interior of the conduit.

Once the conduit is cut, ream it to remove all burrs. Most electricians use a conduit-reaming attachment on a screwdriver. Once conduit is placed in a fitting, you must tighten the setscrew that retains the tubing. If you are not installing a ground wire, a tight connection between the conduit and its fitting is essential. Anchors should be installed on conduit at intervals not to exceed 6 feet. Conduit should be anchored within 2 feet of each box that is served by conduit.

As the diameter of conduit becomes larger, anchors must be closer together. Check your local code requirements for spacing regulations. When conduit will make a turn more than three times before entering a box, you should install a pulling elbow on every fourth turn. Don't splice wire at the pulling elbow. The elbow is meant to provide access only for pulling wire.

Bending Metal Conduit

Bending metal conduit is not difficult, but mastering the techniques can take time and practice. Rookies often crimp conduit, which is a waste of both time and money. Crimped conduit cannot be used. You can avoid bending conduit by using fittings, but this can get expensive on large jobs. A bending tool and some practice will reduce the expense of large jobs.

The most difficult part of bending small conduit is getting the bend exactly where you want it. Let's say that you want to run from one box down a wall, take a 90 degree turn and enter another box. You will have to measure for the bend carefully. Start by measuring from the seated position of the conduit fitting on one of the boxes and measure to the wall that the conduit will run along once it is bent. Take the distance from the fitting to the wall and deduct distance for the bend.

Conduit Bending Distances

✔ A ½ inch conduit requires 5 inches of space for a 90 degree bend.

✔ A ¾ inch conduit requires 6 inches of space for a 90 degree bend.

✔ A one inch conduit requires 8 inches of space for a 90 degree bend.

Assume that you have a measurement of 55 inches from the conduit fitting to the wall. You are working with ½ inch conduit. When you deduct 5 inches for the bend, you need a 50 inch section of conduit to make the bend come out in the right place. Once the conduit-bending tool is on the conduit, you will have to lever it up to bend the tubing. You should use a slow, steady pressure. Jerking on the tool will result in crimping, and that is not acceptable.

Bending offsets in metal conduit is tricky. You will use the stripe painted along the side of the conduit to make offsets. Create a 15

 Trade Tip: If you are installing conduit and will be making more than four turns, install a junction box. The use of junction boxes will ease the pulling of wires.

degree bend. Roll the conduit over, move the bender a few inches farther from the end of the conduit and pull until the section beyond the first bend is parallel to the floor. You may find this difficult at first, but seasoned electricians can make the bends quickly and effortlessly (Figures 5.19 through 5.25).

METAL STUDS

Metal studs require some special provisions for installing electrical cable. Pre-planned holes exist in metal studs to accept electrical wiring. But, you shouldn't just string wire through the holes. Plastic bushings need to be installed before you pull wire. Without the bushings, the sheathing on your cable could be damaged. The bushings come in two pieces and snap together. You could say that installing them is a snap. Bushings must completely encircle the cable that is to be installed.

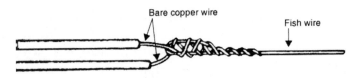

Attachment of stranded conductor to fishing wire

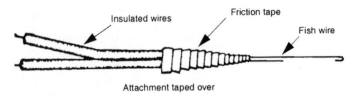

Attachment taped over

FIGURE 5.19 Attachment of fish wire for pulling wire through conduit.

How to Bend a Stub

The stub is the most common bend. Note that your bender is marked with the "take-up" of the arc of the bender shoe.

Example:
Consider making a 14" stub, using a 3/4" EMT conduit.

Step 1.   The IDEAL bender indicates stubs 6" to ↑. Simply *subtract* the take-up, or 6", from the finished stub height. In this case 14" minus 6" = 8".

Step 2.   Mark the conduit 8" from the end.

Step 3.   Line up the *Arrow* on the bender with the mark on the conduit and bend to 90°.

Remember: Heavy Foot pressure is critical to keep the EMT in the bender groove and to prevent kinked conduit.

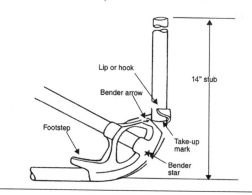

Lip or hook

Bender arrow

Footstep

14" stub

Take-up mark

Bender star

FIGURE 5.20   Techniques for bending conduit. *(Courtesy Ideal Industries, Inc.)*

 Trade Tip: Power benders for conduit are not cheap, but they can save a contractor a lot of money on large jobs. These tools are well worth looking into if you do a lot of bending work.

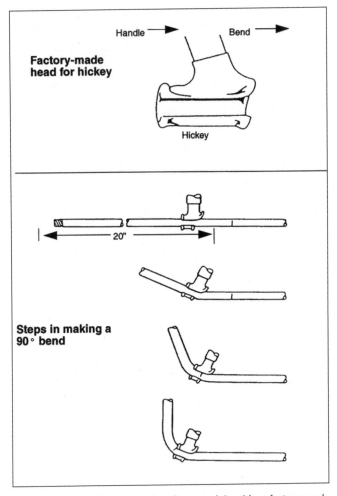

Handle ➜   Bend ➜

Factory-made head for hickey

Hickey

⟵ 20" ⟶

Steps in making a 90° bend

FIGURE 5.21 Technique for bending conduit with a factory-made hickey.

Trade Tip: THHN (90°C small diameter general purpose, 600V building wire), is often used with conduit. The insulation on THHN wire is nylon and is thinner than other types of plastic insulation. This allows more wire to be installed in a conduit.

How to make an Offset Bend

The offset bend is used when an obstruction requires a change in the conduit's plane.

Before making an offset bend, you must choose the most appropriate angles for the offset. Keep in mind that shallow bends make for easier wire pulling, steeper bends conserve space.

You must also consider that the conduit shrinks due to the detour. Remember to ignore the shrink when working away from the obstruction, but be sure to consider it when working into it.

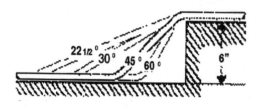

Example:

Step 1.   Measure the distance from the last coupling to the obstruction.

Step 2.   Add the *"shrink amount"* from the table on page 5 to the measured distance and make your first mark. Your second mark will be placed at the *"distance between bends."* (Refer to table on page 5.)

Step 3.   Align the *Arrow* with the first mark and using the *Degree Scale* bend to the chosen angle. Slide down the conduit and rotate conduit 180°, align the *Arrow* and bend as illustrated.

Bend offsets in the air. Remember to keep your body pressure close to the bender.

First Bend

FIGURE 5.22 Techniques for bending conduit. *(Courtesy Ideal Industries, Inc.)*

Second Bend

Example:
30° Bend with a 6″ Offset Depth

Distance Between Bends ← 12″ ⫡ 1-1/2″ → Shrink Amount

Reference Table for Offset Bends

Degree of Bend

Offset Depth Inches	22-1/2°		30°		45°		60°	
2″	5-1/4″	3/8″						
3″	7-3/4″	9/16″	6″	3/4″				
4″	10-1/2″	3/4″	8″	1″				
5″	13″	15/16″	10″	1-1/4″	7″	1-7/8″		
6″	15-1/2″	1-1/8″	12″	1-1/2″	8-1/2″	2-1/4″	7-1/4″	3″
7″	18-1/4″	1-5/16″	14″	1-3/4″	9-3/4″	2-5/8″	8-3/8″	3-1/2″
8″	20-3/4″	1-1/2″	16″	2″	11-1/4″	3″	9-5/8″	4″
9″	23-1/2″	1-3/4″	18″	2-1/4″	12-1/2″	3-3/8″	10-7/8″	4-1/2″
10″	26″	1-7/8″	20″	2-1/2″	14″	3-3/4″	12″	5″

FIGURE 5.23 Reference table for bending conduit. *(Courtesy Ideal Industries, Inc.)*

> ### ⬛ CAUTION
>
> Be sure to line up all bends to be in the same plane.

Hickeys

Hickeys require a different approach to bending. It is not a fixed radious device but rather one that requires several movements per bend. The hickey can give you the advantage of producing bends with a very tight radius.

Order Information

Conduit Size	Ductile Iron Bender	Aluminum Bender	Hickey	Handle
EMT				
1/2"	74-001	74-031	74-010	74-019
3/4"	74-002	74-032	74-011	74-019
1"	74-003	74-033	74-012	74-020
1-1/4"	74-006	74-036	74-013	74-021
Rigid/IMC				
1/2"	74-002	74-032	74-011	74-019
3/4"	74-003	74-033	74-012	74-020
1"	74-006	74-036	74-013	74-021
Handles				
3/4" IPS 38" Long Expanded Extra High Strength Handle				74-019
1" IPS44" Long Extra High Strength Handle				74-020
1-1/4" IPS 54" Long Extra High Strength Handle				74-021

The IDEALbender line gives you the engineering design, indicator marks and durability to bend conduit with ease and confidence.

FIGURE 5.24 Techniques and reference table for bending conduit. *(Courtesy Ideal Industries, Inc.)*

Example:

Step 1. You encounter a 3" O.D. pipe 4 feet from the last coupling. The formula shown in the chart below indicates that for each inch of outside diameter of the obstruction, you must move your center mark ahead 3/16" per inch of obstruction height and make your outer marks 2-1/2" per inch of obstruction height from the center mark.

Step 2. The following table gives the actual mark spacings. In this example, the center mark is moved ahead 9/16" to 48-9/16". The outer marks are 7-1/2" from the center mark, or 41-1/16" and 56-1/16". Mark you conduit at these points.

If Obstruction Is	Move Your Center Mark Ahead	Make Outside Marks From Center Mark
1"	3/16"	2-1/2"
2"	3/8"	5"
3"	9/16"	7-1/2"
4"	3/4"	10"
5"	15/16"	12-1/2"
6"	1-1/8"	15"

Step 3.

(A) Align the center mark with the *Rim Notch* and bend to 45°.

(B) Do not remove the conduit from the bender. Slide the bender down to the next mark and line up with the *Arrow.* Bend to 22-1/2° as indicated.

(C) Remove and reverse the conduit and locate the other remaining mark at the *Arrow.* Bend to 22-1/2° as indicated.

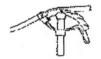

FIGURE 5.25 Techniques and reference table for bending conduit. *(Courtesy Ideal Industries, Inc.)*

SUBPANELS

Subpanels are your friends when you have enough incoming power but don't have enough breaker space in an existing panel. It is common for subpanels to be installed near a main panel, but subpanels are sometimes installed in more remote locations. For example, if you were working with an attic conversion, would you be likely to install the subpanel in the attic?

A double-pole feeder circuit breaker has to be installed in the main panel box to feed power to the subpanel. If you encounter a panel box that doesn't have room for this breaker, you may be able to make room for it by using slimline breakers and reconnecting some of the existing 120 volt circuits to them. Large subpanels can hold up to 20 single-pole breakers.

TIP SHEETS

Start your own tip sheets. When you run into a special situation and solve a problem, write down what was encountered and what you did to overcome the problem. It can be difficult to remember everything. If you keep good notes of what you do in both troubleshooting and installation, you can build your own reference guide. This may seem silly, but it will work. You will be surprised how much data you will collect in just one year. Organize your notes and keep them available for future reference. It will take a little time to record your data, but it may save you a lot of time down the road.

chapter 6

HOME AUTOMATION

The popularity of home automation is growing steadily. There is significant demand for smart homes. Most of the work being done is on new construction. Wiring a smart house is much easier if you can do it as the home is being built. Moving into an existing home and hardwiring it for automation is time consuming and expensive. Even so, people are sacrificing to enjoy the benefits of high tech wizardry.

What's so great about home automation? Convenience and control are two of the primary advantages to having an automated home. With today's technology, a homeowner can control nearly any electrical device from a single controller. This trend started sometime back in the 1970s and continues to grow. As people add new elements, such as computers, to their homes, the desire for having a single control to operate all household appliances and equipment becomes stronger.

Electricians are often called upon to wire a smart house. Sometimes the automation systems in existing homes are achieved without hardwiring. The cost of installing a hardwired system in an existing home can make the process too expensive for homeowners. If you are not up to speed with the options available for installing home automation, you could be losing out on a lot of money.

STRUCTURED WIRING SYSTEM

What is the backbone of a home-automation system? The answer is a structured wiring system. Structured wiring systems use a wide bandwidth that allows more information to pass through the system than can pass through a standard wiring system. This type of wiring system is needed for telephones, fax machines, and high-speed digital computer transmissions. Some of the items that can be controlled with a home-automation system include the following:

- *Telephones*
- *Fax machines*
- *Computers*
- *Electrical devices*
- *Heating units*
- *Air-conditioning units*
- *Plumbing devices*
- *Security systems*
- *Audio equipment*
- *Video equipment*
- *Appliances*

POWER LINE CARRIER

A *power line carrier* (PLC) allows one-way command signals to be transmitted over a standard electrical circuit. There is no need for additional wires when a PLC is used. Digital pulses that are sent through the wires control designated electrical circuits. Many systems are controlled by simple, wall-mounted keypads or touch panel. A keychain remote can also be used to control a system.

Many customers prefer a remote control. If you think about it, why wouldn't they? Do you favor getting up to change television channels, or would you rather change the channels from your chair with a remote control? The same principle applies to home automation.

A keychain remote allows a user to control a home-automation system from their easy chair, their yard, their car, and so forth. This is a big advantage. I think every home-automation system should be set up for remote control capabilities.

TIMERS

Timers have been around for a long time. They have been used for everything from turning on a coffee maker in the morning to controlling lights when people are away from home. Timers from years ago were limited, unfortunately, to operating one device or one power strip. This is not the case with modern timers. Home-automation systems can be outfitted with timers that will operate directly or remotely. A single timer can control multiple devices. Battery backup systems protect the timer from losing control of a system during power failures. High-tech timers offer many possibilities that were not available in the past.

COMPUTERS

Computers can be used to control home-automation systems. Software is available to control devices at specific times and locations with no more effort than clicking a mouse. Some systems are designed to allow remote operation. Wouldn't it be convenient to call your home computer from a jobsite and have the heat turned up to a comfortable temperature when you arrive home from work? With today's equipment, you could. From a phone, it's possible to control any device connected to the automation system.

GOING WIRELESS

Going with a wireless remote control means relying on radio-frequency transmissions. These transmissions operate lights and appliances from a remote location, but distance is a factor. While these units will normally work around the house and yard, they do have a limited range.

Receivers for the remote control may be wall mounted or they may plug into electrical outlets. Existing house wiring is used to conduct the radio signals. This type of system is ideal for an existing home where the cost of a hardwired automation system is prohibitive.

NETWORKING

Most people think of computers when they think of networking, but you can also network equipment in a home. This phase of automation is newer than basic home automation.

New appliances generally come equipped with some type of independently operated electronic processors. When you network a house, the appliances are tied together in a form of communication. This allows appliances to work together to meet your needs. What are some of the items that can be networked? Almost anything can, including the following:

- *Computers*
- *Scanners*
- *Printers*
- *Fax machines*
- *Televisions*
- *DVD players*
- *VCRs*
- *Stereos*
- *Gates*

- *Lights*
- *Switches*
- *Security systems*
- *Irrigation systems*
- *Vents*
- *Heating systems*
- *Air-conditioning systems*
- *Plumbing systems*
- *Pumps*

SURGE CONTROL

Surge control is important with a home-automation system. People are pretty familiar with surge-suppressor outlet strips that are so often used with computers. The surge control for a home-automation system does a similar job, but the design is not nearly the same. A surge control for an automation system fits in the service panel box and takes up no more room than a circuit breaker. The surge control is an essential piece of equipment for an automation system.

Lightning can wreak havoc on an automation system that is not protected with a surge control. Surges and spikes in the automation system can turn the system on automatically, turn it off automatically, or cause the system to malfunction. Having an automation system that is not controlled properly can be both frustrating and dangerous. Don't install a system without a surge control. One surge-control device will completely protect a residential electrical system.

PROTOCOLS

There are four popular protocols used for home automation. Some are more expensive than others. Of the preferred protocols, the X-10 is the least expensive. This protocol transmits over a power line. A CEBus protocol can range from being inexpensive to hitting a middle range of expense. This protocol transmits over a power line, a twisted pair, coaxial cable, radio frequency, infrared, or fiber optics. Obviously, it provides more options than an X-10 protocol, but it can be somewhat more expensive if the options are utilized.

LonWorks is similar in cost to a CEBus system. A LonWorks protocol transmits via power line, twisted pair, radio frequency, or additional types supported by third-party transceivers. A Smart House protocol is considered to be a top-of-the-line protocol. This type of system requires custom wiring to function and its price runs from a mid-range expense to high expense.

X-10 Technology

X-10 technology is ideal for retrofitting existing homes with home automation. Since this protocol controls circuits by sending signals from point to point, there is no need to cut open wires or install raceways to install the system. This is basically the only cost-effective way to fit an existing house with a home-automation system.

The X-10 protocol signals are transmitted by modulating radio-frequency (RF) bursts of 120 kilohertz (kHz) power. Wiring for this type of system is minimal. The RF bursts are made up of a start code, a house code, a function code, and a unit code. These codes run on existing standard electrical wiring in a home. There are 32 address-group codes in an X-10 protocol. Sixteen house codes are common with an X-10 protocol. As a rule, there are also 16 unit codes. This combination allows an X-10 protocol to provide up to 256 unique addresses to assign to individual devices. That's a lot of devices for one home.

Receivers in an X-10 system will respond to one of as many as six commands given for a set address. The six commands are:

- *On*
- *Off*
- *Dim*
- *Bright*
- *All lights on*
- *All lights off*

CEBus Protocol

A CEBus protocol communicates using an electronic command language that is known as Common Application Language (CAL). The language includes device-specific commands like fast-forward, rewind, volume up, temperature down, and so forth. Unlike the X-10 protocol, CEBus uses variable signals to change the intensity of each radio frequency burst. Typically, the CEBus system is more reliable than the X-10 system. This is because the CEBus requires a system to be able to recover from signaling errors. CEBus systems can be transmitted across the following:

- *Power lines*
- *Coaxial cable*
- *Radio frequencies*
- *Low-voltage wires*
- *Category 5 cable*
- *Infrared frequencies*
- *Fiber optics*

LonWorks

The LonWorks protocol is a leader in its field. There are sources that say a LonWorks protocol is the best choice for automated lighting and mechanical systems. This protocol is used in residential properties, but it is also used commercially in transportation, energy management in large buildings, and in industrial equipment control.

Smart House

The Smart House protocol is associated with the National Association of Home Builders (NHAB). Generally speaking, NHAB has invited competing manufacturers to develop products and applications for the Smart House protocol. It is the goal of this program to unify the wiring of systems to work with all forms of automation. Despite the worthy goal, the Smart House system is privately owned and requires custom wiring and service. To some, this is a disadvantage. The cost of using this system can also be a drawback.

HOME SECURITY

Home security is a major consideration when thinking of automating a home. When we talk of security, we are not only talking about bad guys in ski masks. Fire, for example, is a serious threat that a home security system can detect and react to. Some of the functions of a security system include:

- *Deterring burglars and robbers*
- *Detecting home intruders*
- *Detecting approaching vehicles*
- *Detecting smoke*
- *Detecting fire*
- *Detecting gas leaks*
- *Detecting carbon monoxide*
- *Signaling for emergency assistance*

Tying all security features into an automated system makes sense. You have to make sure that the system is as dependable as possible. With proper planning of the right system, you can have exterior lights, interior lights, electrically-operated gates, closed circuit televisions, motion detectors, fire detectors, window-alarm contacts, and other security elements all working together to protect a home.

A security system can sound a local alarm, call a pre-programmed phone number with a recorded message, or do both at the same time. Check manufacturers' instructions for the various components that will be working together.

LIGHTS

Lights are an obvious choice when thinking of what housing components to put on an automated system. It's very convenient to arrive home and only have to click a remote control button to turn on the inside and outside lights. The design of the lighting control can include random on and off switching of interior and exterior lights. This feature would help to deter burglars when the occupants of the house are away. This is also a good safety procedure. Any home that is automated should have the lighting system included on the automated system.

 Trade Tip: If your customer wants a little extra security, consider connecting an interior light to an outside motion detector. If the detector picks up movement, the light will come on inside the building. This can be quite startling to a burglar.

PLUMBING

You probably wouldn't think much about putting plumbing on an automated home system, but maybe you should. Both plumbing and gas piping can be rigged so that the cutoff valves can be controlled by a home automation system. This can be an effective safety feature. For example, you can set up a system to close all gas valves if it detects any conditions indicating fire.

Water heaters can be connected to a system so that the water heater can be shut on and off to conserve electricity when homeowners are away or at work. Pools, spas, and other plumbing-related fixtures can be connected to a system. There are even photoelectric sensors that will turn water on at a faucet when a person's presence is detected. This same type of sensor will flush a toilet when a person moves away from the fixture. Manual overrides and battery backup is common in these applications.

HVAC

HVAC systems can be tied into a home-automation system. If a computer operates a system, a click of the mouse can independently control the temperature in individual rooms. This feature can save a lot of energy. Various temperatures can be pre-programmed into a system to maximize comfort and energy conservation. This, of course, saves the homeowner money for energy that is not used.

Your customer doesn't have to use a computer to gain the benefits of having an HVAC tied into an automated system. Communicating thermostats can get the job done.

Trade Tip: Indoor service panels should be located within 5 feet of the electrical meter. If this is not the case, an outdoor shutoff is required. It makes good sense to install a padlock on the outside cutoff to reduce the risk of pranksters cutting off your electricity.

ENTERTAINMENT

Entertainment equipment in a modern home can be quite comprehensive. No longer is a single radio suitable for home entertainment. Today's homes are usually bulging with entertainment equipment of all types. An automated home system should include entertainment equipment. Some of the types of devices to be connected may include the following:

- *VCRs*
- *DVD players*
- *Digital satellite systems*
- *Interactive television*

- *Standard television*
- *Video games*
- *Stereos*

TELECOMMUNICATIONS

Telecommunications can be networked to enhance a lifestyle. Presently, telecommunications can include:

- *Telephones*
- *Fax machines*
- *Computers*
- *Video exchanges*

- *Message recording*
- *Message retrieval*
- *Teleconferencing*
- *Multiple-line phone systems*

Now that fiber optics are on the scene, other telecommunication options will become available. It may take some time, but you can expect to see the fiber-optic revolution bring new developments that will carry the following:

- *Computer-generated sounds*
- *Video images*
- *Photographic images*

- *Data transmissions*
- *Music*
- *Graphics*

OUTDOOR SYSTEMS

Outdoor systems, such as sprinklers, should be connected to an automated house system. The system can turn a sprinkler system on and off at preset times. A good system gives plenty of options for the efficient use of an irrigation system. Outdoor lighting should be on an automated house system. Anything that uses electricity and is found outdoors is a potential candidate for an automated system.

SYSTEM SPECIFICATIONS

System specifications vary considerably. Most systems are fairly simple to install, especially the X-10 system. You should request installation instructions and specifications from all manufacturers who make the systems you anticipate working with. Read and follow the manufacturer's recommendations. This is the best way to avoid making mistakes by trying to apply generic installation procedures to specific systems.

chapter

7

LIGHTING

S ince the early days of the 20th century, lighting with electricity has been the driving force behind development of electrical power utilization. Compared to open flames in the form of oil or gas lanterns, electric light was relatively safe. Remember how Chicago's great fire started? Well, if Mrs. O'Leary's cow had kicked over an industrial fluorescent strip with a 14 gauge wire guard, Chicago would look slightly different today.

Shortly after Edison designed the first practical incandescent lamp, electricians went to work wiring America for light. Electricians installed wire and light fixtures in homes, offices, barns, hospitals, and schools. Those days have long since passed. Electricians no longer install or maintain light fixtures. Yes, you heard me correctly! Actually, what used to be called a light fixture is now called a luminaire. This new term comes from Article 100 of the 2002 National Electrical Code (N.E.C.). For the purpose of our discussion on lighting in this book, we use the terms "luminaire" and "light fixture" interchangeably.

Today we have several basic types of lighting such as incandescent, fluorescent, high-intensity discharge (HID), and emergency lighting. They may be recessed, surface-mounted, pole-mounted, portable, or even buried in the earth. Lighting can be accomplished through a variety of voltages from low-voltage to high-voltage. Power supplies to fixture lamps can be AC or DC depending on the type of lamp utilized.

As an electrician, you may be called upon by your customers, or your employer, to be their source of lighting information. Having a general knowledge of all types of lighting and expertise in specific areas of lighting will be impressive and leave your customers with a sense of confidence in you. Article 410 of the N.E.C. covers luminaires (lighting fixtures), lamps, and lamp holders. Article 411 covers lighting that operates at no more than thirty volts. The first thing any electrician must do under any circumstances is to become familiar with code requirements concerning a particular application. This is done by studying local and national codes, manufacturers' recommendations, and checking with the code enforcement officer or inspector for local jurisdictions. Luminaires and associated equipment located in hazardous (classified) areas are covered by Articles 500 through 517. Here are some quick code tips on lighting that will be useful in planning and carrying out a project:

- *Live parts must not be exposed to contact under normal operating conditions.*

- *Luminaires in damp or wet locations must be installed so that water cannot enter or accumulate in them.*

- *Fixtures must be listed and marked for use in damp or wet locations.*

- *Incandescent fixtures in closets must have a completely enclosed lamp. Fluorescent fixtures are also acceptable with or without enclosed lamps.*

- *Luminaires must be listed for through-wiring if you intend to extend branch circuit wiring to another fixture.*

- *Luminaires must be designed to provide access to junction points.*

- *Fixtures installed in suspended ceilings must be fastened to the gridwork with bolts, screws, clips, or rivets.*

- *Outdoor lighting fixtures and associated equipment may be supported by trees.*

- *Metal parts of light fixtures must be grounded.*

- *The grounded conductor of a lighting circuit must be connected to the screw shell terminal.*

- *Branch circuit conductors within 3 inches of a ballast must be rated 90 degrees C.*

- *Recessed incandescent fixtures must be thermally protected.*

- *Only type IC rated recessed fixtures are permitted in direct contact with insulation and combustible material. Non-type IC fixtures must be kept 3 inches away from these materials.*

- *Luminaires installed on DC circuits must be marked for DC operation.*
- *Discharge lighting in dwellings is limited to 1000 volts open circuit.*
- *Lighting track must not extend through a wall or be installed in damp or wet locations.*
- *Lighting track must be properly grounded and securely fastened in place at intervals of 4 feet or less.*
- *Systems operating at 30 volts or less must be listed for the use intended.*
- *Branch circuits must not exceed 20 amperes for low voltage luminaires.*

INCANDESCENT LIGHTING

Thomas Edison's ingenuity and perseverance led to the development of the incandescent lamp, or bulb as it is commonly referred to, in 1879. Since that time lamp types have been greatly improved. Incandescent lamps are extremely popular in residential wiring, due mainly to the lower cost of lamps and luminaires. Ease and simplicity of maintenance by the homeowner further adds to the popularity of this type of lighting. The lamp consists of a tungsten wire filament supported in a glass bulb (Figure 7.1). The air inside the bulb has been removed and replaced with argon or another inert gas. One end of the filament is connected to the center contact point on the base of the bulb; the other end is connected to the screw shell. When a typical light bulb is screwed into a light socket, the hot wire connects to the center contact while the neutral wire connects to the screw shell. When 120 volts of power is first applied, the circuit sees it as a short circuit. It's almost like touching the black and white wire together. Of course, any electrician will tell you that if you did this you would trip the circuit breaker or blow the fuse. The reason you don't trip the circuit when an incandescent lamp is energized is that as the filament gets hot, it limits the flow of electric current to the wattage rating of the lamp. Since wattage equals volts times amps, a 60 watt lamp at 120 volts draws 0.5 amps. Example: 60 watts = 120 volts x 0.5 amps. From this equation you can see why this action would not trip a 15 or 20 amp circuit.

Have you ever noticed that when light bulbs blow, it is usually when they are first turned on? That's because the filament hasn't quite heated up yet and current hits it like a dead short. In fact, if an incandescent lamp is tested with a continuity tester it will read as a

Inert Gas

Filament

Screw Shell

Center Contact

FIGURE 7.1   Incandescent light bulb.

dead short. This must be taken into consideration whenever trouble shooting a circuit. If, for example, the circuit has a short in it and you are trying to isolate the short with a tester, any connected lamps will interfere with your testing. The best way to avoid confusion is to remove lamps from the circuit being tested. Incandescent lamps are represented by a number that is the diameter of the lamp in eighths of an inch. For example an A19 lamp, which is a typical light bulb, has an arbitrary shape and is 19/8 inches in diameter. That is to say, the bulb is 2⅜ inches in diameter. Of the three main classifications of lighting (incandescent, fluorescent, and HID), incandescent is the least efficient.

FLUORESCENT LIGHTING

The first fluorescent lamp was developed in 1896. The N.E.C. classifies fluorescent fixtures or luminaires as electric discharge lighting. Fluorescent lighting carries the advantages of long life and high efficiency. Lamp replacement is relatively simple for the homeowner. Luminaires in this class usually run at much lower operating temperatures than do those in the other two classes. A fluorescent lamp is a glass tube with an electrode at each end. The tube is coated inside with phosphor and filled with inert gas. A small quantity of mercury is released into the tube also. Fluorescent light is produced as the electrodes arc at each end through the vapor inside the tube (Figure 7.2). As the atoms in the vapor become more and more ionized, the resistance to current flow is lowered. Without being limited, this process would continue to grow until the lamp is destroyed. Arc current must be limited and controlled. This is done by the ballast, which is actually wired in series with the lamp. Ballasts installed inside must be thermally protected and will be designated as a Class P ballast. However, Class P ballasts are not permitted for exit fixtures or for fixtures energized only during times of egress.

The latest technology in fluorescent lighting is the widespread use of electronic ballasts. The power factor of an electronic ballast can be as much as .95 or more. Fluorescent lamps are about five times more efficient than incandescent lamps at converting electricity. Lamp life is typically 10,000 to 20,000 hours. Electronic ballasts consume approximately 25 percent less wattage than the standard magnetic ballasts. In addition, electronic ballasts are quieter, lighter in weight, and eliminate lamp flicker. These are great selling points. Everybody likes to save money; therefore, the 25 percent reduction in power consumption will get anyone's attention. In my experience as a contractor over the years, the two biggest gripes people have with fluorescent lighting are the grinding noise produced by ballasts and flickering tubes. These two characteristics associated with the older type magnetic ballasts are not inherent to the newer electronic type (Figure 7.2).

Fluorescent lamps, or tubes as they are more commonly known, are represented by a number that is the diameter of the tube in

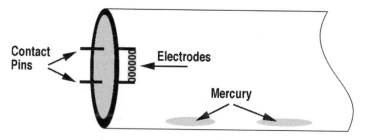

FIGURE 7.2 Cutaway of a fluorescent tube.

fastfacts

Tubes for fluorescent lamps often contain mercury. Never allow a fluorescent tube to break. Check with your local authorities for proper disposal guidelines.

eighths of an inch. Older style tubes used in luminaires with magnetic ballasts are T12; which means they are 12/8 of an inch in diameter. This translates to one and a half inches. The latest style of electronic ballast luminaires uses a T8 lamp. These tubes are 1 inch in diameter.

It is our policy in business to always push for replacing T12 luminaires with T8 luminaires. Your customers will never be disappointed. It is a step into the 21st century and customer satisfaction and power efficiency are guaranteed.

HIGH INTENSITY DISCHARGE (HID)

The first mercury lamp was produced by Peter Hewitt in 1901. This was the beginning of HID lighting. HID lighting is similar to fluorescent lighting, which is also classed as discharge lighting by the N.E.C. Today, HID lighting has evolved into three subclasses: mercury vapor, metal halide, and high-pressure sodium. Each class must have a ballast to limit the current. HID lamps are characterized as negative resistance light sources. If the ballast weren't there to limit the current, then the lamp would quickly destroy itself. A transformer integral to the ballast circuit matches the required lamp voltage with the available circuit voltage (Figures 7.3 through 7.13). The HID ballast serves four basic functions:

1. Matches the line voltage to the required lamp operating voltage.
2. Regulates the input voltage to provide proper lamp lumen output.
3. Limits the allowable lamp current, since all ballasted lamps have a negative resistance.
4. Provides the required open circuit voltage to start the lamp.

HIGH PRESSURE SODIUM

Ballast Type	Available Input Voltage	Starting Current (Power Factor)	% Line Variation= % Wattage Change	Ballast Losses	Lamp Current Crest Factor
REACTOR 	120, 50, and 150W lamps.	Higher than operating (50% NPF Standard) (90% HPF Available)	± 5% = ± 12%	LOW	1.4 to 1.5
HIGH REACTANCE AUTOTRANSFORMER 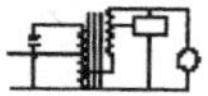	All voltages for 70, 100 and 150W lamps.	Slightly higher than operating for 100 and 150 watt. Less than operating for 70 watt. (90% + HPF)	± 5% = ± 12%	MEDIUM TO HIGH	1.5
CONSTANT-WATTAGE AUTOTRANSFORMER (AUTO-REGULATED LEAD) 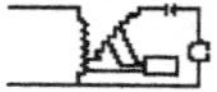	All voltages for 200 250, 310, 400 and 1000W lamps	Less than operating (90% + HPF)	± 10% = ± 10%	MEDIUM TO HIGH	1.5
CONSTANT-WATTAGE (MAGNETIC REGULATOR OR REGULATED LAG) 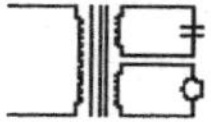	All voltages for 200 and 400W lamps.	Less than operating (90% + HPF)	± 10% = ± 3%	HIGH	1.7

FIGURE 7.3 High pressure sodium ballast circuits. (*Courtesy Lithonia Lighting*)

METAL HALIDE

Ballast Type	Available Input Voltage	Starting Current (Power Factor)	% Line Variation= % Wattage Change	Ballast Losses	Lamp Current Crest Factor
HIGH REACTANCE AUTOTRANSFORMER (HX)					
	All voltages except 480V for 70, 100 and 150W lamps.	70W is slighly lower than operating 100W is slightly higher than operating (90% + HPF)	± 5% = ± 10%	MEDIUM	1.5
CONSTANT-WATTAGE AUTOTRANSFORMER (PEAK-LEAD)					
	All voltages for 175W lamps and higher.	Lower than operating (90% + HPF)	± 10% = ± 10%	MEDIUM	1.6 to 1.8
LOW LOSS LINEAR REACTOR BALLAST (LLRPSL--PULSE START)					
	277V only for 150-450W lamps.	Higher than operating (90% + HPF)	± 5% = ± 10%	LOW	1.45
SUPER CONSTANT-WATTAGE AUTOTRANSFORMER (SCWA--PULSE START)					
	All voltages for 150W lamps and higher.	Lower than operating (90% + HPF)	± 10% = ± 10%	MEDIUM	1.60

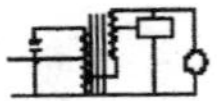

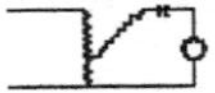

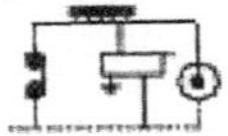

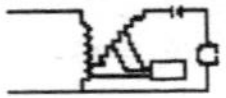

FIGURE 7.4 Metal halide ballast circuits. (*Courtesy Lithonia Lighting*)

MERCURY

Ballast Type	Available Input Voltage	Starting Current (Power Factor)	% Line Variation= % Wattage Change	Ballast Losses	Lamp Current Crest Factor
REACTOR					
	240 & 277V for 100, 175, 250 and 400W. 480V for 1000W lamps.	Higher than operating (50% NPF Standard) (90% HPF Available)	± 10% = ± 5%	LOW	1.4 to 1.5
CONSTANT-WATTAGE AUTOTRANSFORMER					
	All voltages for all lamp wattages.	Lower than operating (90% + HPF)	± 10% = ± 5%	MEDIUM	1.6 to 2.0
CONSTANT-WATTAGE AUTOTRANSFORMER (AUTO-REGULATED LEAD)					
	All voltages for all lamp wattages.	Less than operating (90% + HPF)	± 13% = ± 2%	HIGH	1.8 to 2.0

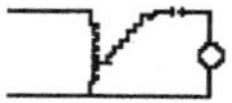

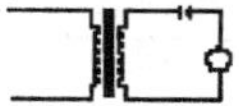

NOTE: Ungrounded power distribution systems may carry transient line voltage under fault conditions. Because high transients can cause premature ballast lamp failure, it is not recommended that luminaires be operated on any ungrounded 480V or other ungrounded system.

FIGURE 7.5 Mercury ballast circuits. *(Courtesy Lithonia Lighting)*

	LAMP		Open Circuit Voltage RMS	Secondary Short Circuit Current Amps
	Wattage	ANSI Number		
MERCURY BALLASTS	50	H46	225–255	0.85–1.15
	75	H43	225–255	0.95–1.70
	100	H38	225–255	1.10–2.00
	175	H39	225–255	2.00–3.60
	250	H37	225–255	3.00–3.80
	400	H33	225–255	4.40–7.90
	2–400 (ILO)	2–H33	225–255	4.40–7.90
	2–400 (Series)	2–H33	475–525	4.20–5.40
	700	H35	405–455	3.90–5.85
	1000	H36	405–455	5.70–9.00
METAL HALIDE BALLASTS	High Reactance Autotransformer (HX)			
	70	M98	230–280	0.95–1.25
	100	M90	240–275	1.35–1.70
	150	M102	235–290	2.05-2.55
	Constant Wattage Autotransformer (CWA)			
	175	M57	285–320	1.50–1.90
	250	M80	230–270	2.90–4.30
	250	M58	285–320	2.20–2.85
	400	M59	285–320	3.50–4.50
	2–400 (ILO)	2–M59	285–320	3.50–4.50
	2–400 (Series)	2–M59	600–665	3.30–4.30
	1000	M47	400–445	4.80–6.15
	1500	M48	400–445	7.40–9.60
	Low Loss Linear Reactor Pulse Start (LLRPSL)			
	150	M102	250-305	2.00-2.50
	175	M137	250-305	1.70-2.10
	200	M136	250-305	1.80-2.70
	250	M138	250-305	2.40-3.00
	320	M132	250-305	3.00-3.70
	350	M131	250-305	3.40-4.40
	400	M135	250-305	3.70-4.50
	450	M144	250-305	4.20-5.20
	Super Constant Wattage Autotransformer Pulse Start (SCWA)			
	150	M102	215-265	2.15-2.65
	175	M137	240–290	1.85-2.25
	200	M136	215-265	1.90-2.30
	250	M138	240–290	2.35-2.90
	320	M132	240–290	3.00-3.70
	350	M131	240–300	3.00-3.75
	400	M135	240–300	3.60-4.40
	450	M144	255-315	3.90-4.75
HIGH PRESSURE SODIUM BALLASTS*	35	S76	110–130	0.85–1.45
	50	S68	110–130	1.50–2.30
	70	S62	110–130	1.60–2.90
	100	S54	110–130	2.45–3.80
	150	S55	110–130	3.50–5.40
	150	S56	200–250	2.00–3.00
	200	S66	200–230	2.50–3.70
	250	S50	175–225	3.00–5.30
	310	S67	155–190	3.80–5.70
	400	S51	175–225	5.00–7.60
	1000	S52	420–480	5.50–8.10
LOW PRESSURE SODIUM BALLASTS	18	L69	300–325	0.30–0.40
	35	L70	455–505	0.52–0.78
	55	L71	455–505	0.52–0.78
	90	L72	455–525	0.80–1.20
	135	L73	645–715	0.80–1.20
	180	L74	645–715	0.80–1.20

FIGURE 7.6 HID ballast testing. *(Courtesy Lithonia Lighting)*

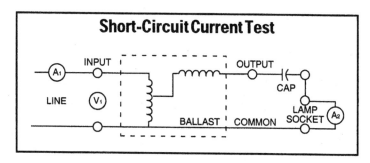

FIGURE 7.7 HID short circuit current test. *(Courtesy Lithonia Lighting)*

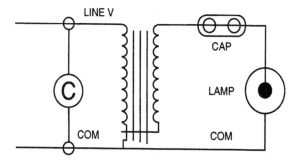

FIGURE 7.8 HID ballast continuity between common and line leads. *(Courtesy Lithonia Lighting)*

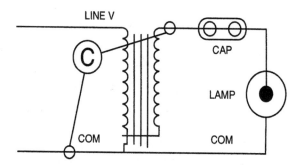

FIGURE 7.9 HID ballast continuity between common and capacitor leads. *(Courtesy Lithonia Lighting)*

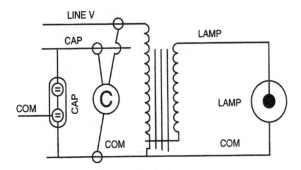

FIGURE 7.10 HID ballast continuity between line and lamp leads. *(Courtesy Lithonia Lighting)*

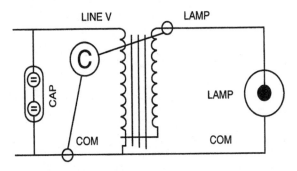

FIGURE 7.11 HID ballast continuity between common and lamp leads. *(Courtesy Lithonia Lighting)*

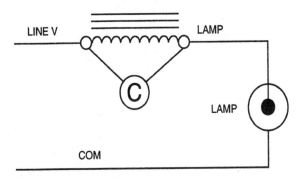

FIGURE 7.12 HID ballast continuity between common and capacitor leads. *(Courtesy Lithonia Lighting)*

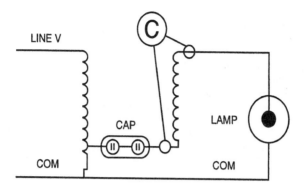

FIGURE 7.13 HID ballast continuity between common and lamp leads. *(Courtesy Lithonia Lighting)*

MERCURY VAPOR LAMPS

Mercury vapor lamps emit a blue green light which is suitable for outdoor use. Average lamp life is about 16,000 hours. Mercury vapor luminaires will work efficiently in extremely cold weather, which make them well suited for street lighting in northern climates. There are hazards involved with mercury vapor lamps, however. If the outer envelope of the lamp is broken or punctured and the arc tube continues to operate, short wave radiation (ultraviolet) will be produced. This can cause serious skin burns and eye inflammation. There are lamps on the market that will automatically shut down if the outer envelope is broken.

METAL HALIDE LAMPS

Metal halide lamps are similar to mercury vapor as far as the hazards are concerned, however, lamps are available that will automatically extinguish when the outer envelope is broken. Metal halide produces a clean white light with a higher efficiency rating than mercury vapor. Lamp life is shorter, about 6,000 hours, and the rate of output depreciation over its lifetime is high.

HIGH PRESSURE SODIUM LUMINAIRES

High pressure sodium (HPS) luminaires are the most energy efficient of HID lighting available. Average lamp life comes in at 24,000 hours. Starting temperatures for these fixtures dip to 40 degrees below zero. HPS lamps don't carry the hazard warning associated with mercury vapor and metal halide lamps. The only drawback to HPS lighting is the color. HPS luminaires emit yellowish light. Sometimes it may even appear purplish. HPS lighting is best suited as perimeter security lighting around buildings, but the deviance from white light makes it unsuitable for most interior applications.

RESIDENTIAL LIGHTING

When it comes to residential lighting, it's all about convenience. Article 210 of the N.E.C. calls for lighting outlets to be installed in specific areas. The number one thing that an electrician must realize about the N.E.C. is that it is a book containing the rock bottom minimum requirements for electrical installations. In the introduction in Article 90, the N.E.C. makes it clear that it is not a design specification nor is it an instruction manual for untrained personnel. It also makes it clear that it is not a guidebook for an efficient, convenient electrical system nor does it provide for good service or future expansion of electrical use. That is why your role as the electrician is so important. You are a professional tradesman proficient in the installation of electrical equipment. You know the codes but beyond that, you also must know what is best for each application.

Basically, what Article 210 tells you is that a wall switch controlled lighting outlet must be installed in each livable room and bathroom. Lights must also be provided for stairways, hallways, garages, exterior entrances, and storage spaces. Lighting must be provided for any equipment requiring servicing.

Article 210 by exception allows you to wire for one or more switched receptacles in any room other than a kitchen or bathroom.

 Trade Tip: If you are installing boxes in an unfinished area, such as a garage, basement, or attic, you should use metal boxes. Leaving plastic boxes exposed can result in damaged boxes.

fastfacts

If you find a light switch that has white neutral wires connected to the switch, watch out. There may be power going to the light fixture even when the switch is off. The black, hot, wires should be connected to the switch. When working on older electrical systems, never assume that the light switch cuts off the ungrounded conductor. Turn the power off at the source.

This arrangement is designed for people who, for whatever reason, don't want a ceiling light in a room. The wall switch controlled receptacles are there to energize a table lamp or a floor lamp when the receptacle is turned on. The downside to this method is that someone might turn the lamp off by its built in switch, not knowing about the wall switch controlled receptacle. The next person comes along, turns the wall switch on and off several times with no success. Then he or she goes to the lamp switch and turns it on and off several times with no success because the wall switch was left in the off position. Now this person is completely frustrated and left standing in the dark wondering why the lamp won't work. The switched receptacle works perfectly only if the user fully understands how it works. Wall switch controlled ceiling luminaires usually come with much less confusion. That is why we usually plan for ceiling lights in the habitable rooms of a house.

Now that we have a good understanding of what the N.E.C. requires for residential lighting, let's plan out the lighting system for a new home. We will be lighting a two story colonial with a full unfinished basement. The structure is approximately 1200 square feet per floor. We will begin in the basement and work up through each floor and on into the attic if necessary.

Basement Lighting

Lighting in unfinished basements is very straightforward. First, check with the owner and get an idea of what the basement will be used for. In many cases, the entire basement is nothing more than a large utility room, a place for the oil tank, heating system, water heater, water softener, and electrical service panel. In such instances, the basement area is generally not visited on a regular basis. In fact, the only time anyone goes into the basement is to work on equipment located there or maybe to store a box or two of something that should have been thrown away.

This basement will be lighted by six incandescent lampholders with 100 watt lamps installed. All six will come on together and off together. The wall switch will be at the head of the cellar stairs just inside the door and on the strike side of the door. In addition to the six luminaires in the basement, one additional luminaire will be located in the stairway to light the steps. The lampholders in the basement should be laid out as symmetrically as possible. Draw out the area on a piece of paper. Now divide the area into six equal sections, two wide and three long. Next, find the center of each section. This layout will give the most even light without having any bright or dark spots.

This layout looks great on paper, but some of the lamp holders may need to move one way or the other for fit in around all the equipment, piping, chimney, and ductwork that is in the basement. When the job is complete and the basement lights are on, make sure any equipment needing routine servicing is adequately lit. To light the stairs, a simple incandescent wall light, such as a jelly jar fixture, will be adequate.

This is the easy basement to light. Basement lighting can get a bit more complex depending on the way the homeowner plans to use it. Sometimes the basement is divided and having all luminaires on or off together is impractical and inconvenient. There may be a small workbench area with separately switched fixtures. There may be a storage area with a separately switched fixture. It is not uncommon to have laundry facilities in an unfinished basement. Some of these areas might be better served with fluorescent fixtures. These fixtures would be controlled in their general area. It is always wise to have

fastfacts

A pull chain fixture can be modified to work with a wall switch by running a two wire cable from the fixture to a switch box. The wires should be connected for end line wiring.

fastfacts

Circuit needs for general living space, such as hallways, bedrooms, and so forth, are often computed based on square footage. A rule of thumb is to install one 15 amp circuit for every 500 square feet of living space. Lighting should be installed on a separate circuit.

the light in the stairway and at least one more light at the bottom of the stairs controlled together. The two stairway lights should be on a set of three way switches, one switch at the top of the stairs and one switch at the bottom of the stairs. This way, the stairway lights won't be on the entire time the homeowner is doing laundry or working on a project on the workbench. Once he or she gets to the bottom of the stairway and turns the light on in the area he or she will be working, the stairway lights can be turned off. If the basement has an exterior entrance, you will want to have a switch near it as well. This switch should be a three way, paired up with another three way somewhere inside. They would need to control enough lighting to allow a person to safely walk through the basement without bumping into anything in the dark.

Now that we have the basement lights all set up with switches conveniently located for the homeowner, we're ready to take our tools upstairs and begin.

Lighting for Living Space

Any room with one entry point is simple. The switch box goes near the door opening inside the room and on the strike side of the door. If the room has no door, then the switch box could go on either side of the opening. A good mounting height for light switches is somewhere around 48 inches. As a rule we like to place ours at 48 inches to the top of the box. Occasionally, however, this height interferes with architectural elements in the room such as chair rails, raised paneling, brick or tile. In these situations the homeowners' personal preference will dictate the mounting heights.

The next question that comes to mind is how many switches will be located at this point. If there is only one light in this room then one switch will be adequate; but there may be a ceiling fixture, wall sconces, and a ceiling paddle fan. Now we're looking at three or more switches. Generally speaking, you should always gang switches, including dimmers and speed controls, together and have one multi-gang wall cover plate over them. Be careful if you're ganging dimmers together. They emit heat under normal operating conditions and many of them require more than one gang per dimmer. For example, you may need a three gang box for two 1000 watt dimmer switches. If a room has more than one entrance, then a switch should be set next to each opening.

Halls and Stairways

Luminaires in halls and stairways should have wall switches conveniently located at each entry point to the area. This house has a large hall area downstairs with one ceiling fixture in the center. The ceiling light is controlled by a switch at the bottom of the stairs, a switch at the front door leading into the hall, a switch where the hall leads to the kitchen, and a switch where the hall leads to the living room. That's one light controlled from four locations. It has to be; otherwise the residents would have to walk through a dark area to get from one room to the other. The switching in this circuit is accomplished by using two three-way switches and two four-way switches. 12/3 cable is run from switch to switch to make the system work. Many times, in large hallways like this with several entry points, one switch can be placed where it can be conveniently operated from two or three entrance points, such as between two adjacent doors. With this general idea of how to lay out wall switches, let's look at each room of the house specifically.

Kitchen Lighting

The kitchen is often the busiest room in the house. Not only is food prepared here, but many activities such as eating, entertaining, homework, reading, and just relaxing may take place in this area. Light levels in the kitchen should be the highest in the house. Decorative fluorescent fixtures mounted above the center of a workspace are a good choice. Kitchens less than 100 square feet require a two lamp fluorescent; four lamp fixtures will light up to 250 square feet. Larger kitchens will require supplemental lighting. Usually there is a light in the exhaust hood over the range. Many homes now install a combination microwave/exhaust hood/light unit over the

fastfacts

Lights in halls and stairways are generally required to be controlled by three-way wall switches at each entrance point to the area in question.

range. These combination units plug into a dedicated branch circuit receptacle in the wall cabinet above the unit. Control of the light is done by a built in turn switch. Standard range hoods have built in light switches as well.

Recessed down lights mounted 12 to 18 inches off the edge of the cabinets and spaced three to four feet from center to center are an excellent way to create additional general lighting. Installing fluorescent under cabinet fixtures help to reduce shadows and add critical light to the workspace. These fixtures operate very cool and are a cost efficient lighting source. In open areas over sinks, use recessed down lights mounted directly over the sink. Switching for these fixtures is usually done locally. That is, recessed down lighting over the counter top should be controlled by a switch near the counter top. Under cabinet lighting can be controlled by a wall switch or by integral rocker switches built into the light. The light over the kitchen sink should be switched to the right or left of the sink.

When installing fluorescent fixtures you should always use luminaires with electronic ballasts and T8 lamps. When installing recessed down lights you should always use type IC so that you don't have to worry about contact with insulation or combustibles. Another nice touch in the kitchen is to control incandescent down lights with dimmer switches. At night they can be set to a low glow and serve as nightlights for anyone browsing around for a snack or returning dirty dishes to the kitchen sink.

 Don't Do This! Don't skimp on kitchen lighting. Kitchens should be well equipped with lighting. At a minimum, the overhead lighting in a kitchen should be either two 40 watt fluorescent lights, two 150 watt incandescent lamps, or four 100 watt incandescent lamps.

Trade Tip: When wiring a kitchen, consider offering your customer the option of cove lighting. This is lighting that is installed on top of wall cabinets. The lighting gives the kitchen a distinctive look when the lights are on.

Ceiling paddle fans are becoming more and more popular these days as another option for a central ceiling fixture in the kitchen. Whether the central ceiling fixture is a decorative fluorescent fixture, an incandescent ceiling light, or a paddle fan with light kit, it should be controlled at each entrance to the room. When someone walks into the kitchen the light can be turned on; when the person leaves the kitchen the light can be turned off. If a person is working in the kitchen, the lighting in a specific area can be turned on in that area and off again when the task is completed.

Dining Room

Lighting in the dining room focuses around the table. A chandelier or pendant can be a general lighting element as well as the focal point of the home. Recessed wall washers can provide additional light while helping to create an illusion of a larger space. A chandelier should be 6 to 12 inches smaller than the narrowest side of the table. The bottom of a chandelier or pendant should be approximately 30 inches above the table. Adjustable recessed fixtures aimed on the table and chandelier helps to provide additional light on the table and will also bring out the brilliance of the chandelier.

Wall sconces placed on either side of doors, china cabinets, or hutches also adds a touch of elegance in this most formal room. We recommend controlling all the lighting with dimmers. That way, just the right level of light can be set for any situation. If the dining room has more than one entry point, we would only control one light with three ways and four ways, if necessary, at each entry point. This would ensure a safely lit path for anyone walking through the room. The other lights, used mostly for accent lighting, would be controlled from a single convenient location.

When roughing in the wiring for a dining room chandelier, it is wise to install a 2 x 6 or 2 x 8 block above the ceiling. Lay it flat and secure it between two floor joists. Then mount an octagon box to it. This will give you something solid to support the chandelier to. Remember out-

fastfacts

What is the minimum circuit requirement for a kitchen? Always check your local code requirements, but, normally, two 20 amp circuits are required. However, the circuit must be rated as a ground fault interrupter circuit (GFI) on receptacle outlets located within 6 feet horizontally of the sink. Additionally, a 240 volt circuit is required for an electric range. Separate circuits should be installed for all major appliances, such as refrigerators, dishwashers, microwaves, and trash compactors. These circuits should be sized based on the wattage ratings of the appliances to be used. Lighting should be installed on a separate circuit.

let boxes can support light fixtures only up to 50 pounds and many large ornate chandeliers would exceed that weight.

Living Room, Family Room, and Den

Lighting in these rooms is used for general illumination. If the room is generally square, one central fixture will work. If the room is more rectangular, two ceiling fixtures will provide more even lighting in the room. Divide the room area into two equal sections. Next, find the center of each half and locate the light at that point.

If ceiling fans with light kits are used, place a light switch at each entry point into the living room. The fans usually are controlled by a speed switch located with the light switch at the entry point that is most often accessed.

 Don't Do This! Don't installed recessed lighting fixtures in a way that will put the housing within 3 inches of insulation. Unless the fixture carries a rating that allows closer contact with flammable materials, keep the housing at least 3 inches from any object that might be ignited or melted from the heat transferred from the fixture housing.

Did You Know: that sconce lighting should be installed at a height of 72 to 78 inches above the floor? This height is recommended to avoid human contact with the fixtures.

Lighting in these rooms can also be used to enhance the atmosphere of the room. Recessed fixtures or track fixtures help to make rooms come alive by dramatizing wall textures or highlighting artwork. Visual tasks such as reading, playing games, or hobbies require more light than would be necessary for general socializing or relaxing. This lighting could be provided by recessed down lighting or pendants located more toward the perimeter and corners of the room.

Track lighting is another good choice for providing task lighting. This system is capable of accepting multiple luminaires simply by snapping them into place. Luminaires in lighting track can be moved, added, or removed any time the use or furniture layout of the room changes. Also, most track luminaires are fully adjustable so that aiming them in any direction is possible.

These rooms are the areas where some people prefer not to use ceiling fixtures. Wall sconces can be laid out to light the sides of entry points. They will also look great to the sides of fireplace mantels or large pieces of furniture. Portable lamps, operated by wall switch controlled receptacles, would be an acceptable alternative to ceiling lights for these rooms.

Halls, Foyers, and Stairways

The foyer conveys the first impression of the interior of the home. It is often viewed from the exterior and makes a transition to the rest of the home. Size the decorative fixture to the space. Two story foyers will require a larger fixture. Make sure you use a fixture that will look attractive from the second floor, if it can be seen from that level.

 Don't Do This! Plastic boxes are very common and are terrific when used for protected indoor wiring. However, don't use plastic boxes for heavy light fixtures or fans.

fastfacts

Wall sconce lighting is a good way to make a room appear larger. The indirect light is normally placed higher than eye level, 72 to 78 inches above the floor. Bulb wattage should be low for this type of lighting.

Stairways and halls must have good illumination for safe travel through these areas. Use matching close-to-ceiling fixtures for hall-ways and smaller chain hung fixtures for stairways. Ceiling fans with light kits are another excellent choice for two story foyers. The fan can be used to circulate air during summer and to blow heat down in the winter.

In addition to the central fixture in the foyer, matching wall sconces can be placed near the landings. Sconces should be mounted above eye level so that the light source is not visible. This usually falls in the range of 72 to 78 inches above the floor level. Make sure you place a wall switch near every entry point to these areas. It is not uncommon to have three, four, or more switch locations for these lights.

Bathrooms

When you are planning the lighting in a bathroom, the key is to provide for shadow free lighting around the mirror. One fixture mounted over the mirror is a good way of lighting the bathroom, but can cause shadows on the face. The addition of recessed down lights mounted 1 to 3 feet on centers between the face and the mirror is a better method for lighting the face and head. Adding wall sconces to the side of the mirror is the best way to be sure there will be no shadows on the face. If the fixtures use exposed lamps, use no higher than a 40 watt lamp. For fixtures with diffused glass you can go up to 75 watts. If fluorescent lighting is desired, make sure you use color corrected lamps.

If you install a recessed fixture in the shower it must be listed and approved for the use. Also, this luminaire must be protected by a GFI circuit. The most convenient way of accomplishing this is to feed the light switch with a wire taken from the load side terminals of the GFI receptacle near the lavatory. This way, should the GFI trip, it can be

fastfacts

Recessed lighting can generate a lot of heat in the surrounding area. This type of lighting needs room to breathe. It should not be smothered in insulation. Avoid leaving flammable materials in the immediate area of a recessed light fixture. The smartest way to end this hazard is to always use recessed fixtures that are designed and listed for direct contact with insulation and combustible materials.

reset in the bathroom; whereas, if a GFCI breaker were used, the homeowner would have to go to the circuit breaker panel to reset the light in the shower.

Another welcome addition to any bathroom is a ceiling exhaust fan. You must vent the fan directly outside. This can usually be done with a section of flexible metal duct connected to a dryer vent kit mounted in the exterior wall of the bathroom. The primary purpose of the exhaust fan is to remove moisture, in the form of steam, from the bathroom. The secondary purpose of the fan is the removal of foul odors from the room. Bathroom exhaust fans can be purchased with a built in light, night light, electric heater, or any combination of the three. A combination fan and light located in the center of the ceiling provides great general illumination for most tasks in this room. Separate switch control of each feature of a combination unit is recommended. This way, the occupants of the home do not have to run the fan if it is not necessary; likewise, the fan can be used without the light if it is not needed. Larger units with built in electric heaters need a dedicated circuit run to them. Units placed in the shower or over the bathtub must be protected by a GFI circuit.

 Did You Know: that rotary snap switches were used from around 1900 to around 1920? These switches were then replaced by push button switches. The push button switches were used until around 1940, when toggle switches took their place.

Bedrooms

Most bedrooms are typically lit in the same fashion. You have one ceiling fixture located in the center of the room. It is controlled by a wall switch mounted inside the room near the strike side of the door. If you want to get a bit fancier, use a dimmer switch. This way, the fixture becomes a night light for small children or anybody who gets up frequently in the night.

Ceiling fans with light kits are very popular in bedrooms today. Control is made from a two gang box near the door. One gang can hold a switch or dimmer for the light while the other gang holds a switch or speed control for the fan. I have this arrangement in the four bedrooms of my house. We use the ceiling fans year around. The rotary dimmers that control the lights naturally extend lamp life by warming up the filaments slowly. Remember from the preceding text that when an incandescent filament is first energized, it feels the inrush like a dead short until it heats up and limits current flow to the rated lamp wattage. Well, a rotary dimmer applies power gradually to the lamp filament, thus extending lamp life.

Some homeowners like to get a bit fancier with the lighting in the master bedroom. Recessed down lights or wall washers can be used to highlight artwork, furniture, or vanity area. Wall sconces can be used for many of the same purposes. In larger bedrooms, or any bedroom for that matter, it is very convenient to wire for a three way switch adjacent to the headboard of the bed. The down side to doing this is that the bed pretty much has to stay in that location permanently. That is usually no problem because bedrooms are designed for basically one furniture layout.

Of course, all this lighting sounds great, but you will always have customers who do not want ceiling lights or wall lights. The N.E.C. allows you to provide switch controlled receptacles for the connection of portable lamps. Which receptacle do you switch? Well, why not switch them all just to be safe? Actually, whenever we talk about

fastfacts

A receptacle, with GFI protection, is required within 3 feet of each basin in a bathroom. The receptacles must not be mounted face up in the countertop.

fastfacts

*When you are wiring a room that does not have a switch oper-
ated, built in light, you must install a switched outlet for the use
of a lamp.*

switching a receptacle we mean only half of a duplex unit. This is
done by running three conductor cable from box to box around the
room. Usually the black conductor will be hot all of the time while
the red conductor will be on or off with the wall switch. When you
wire receptacles in this fashion, you must break the jumper tab off
from between the two brass colored terminals. This is best accomplished by working it back and forth with a small pair of needle
nosed pliers. Black connects to one screw and red connects to the
other. All receptacles in the room should be wired uniformly so that
all of the top receptacles are controlled by the wall switch. Wiring a
duplex receptacle in this fashion is called split wiring. The outlet is
known as a split wired receptacle.

Closets

Lighting inside a closet is optional. In my opinion, it adds a great
deal of convenience for the occupants, not to mention adding value
to the house. Closet lighting can be done most simply by placing a
wall switch inside the closet on the strike side of the door. Connect
the switch to an incandescent jelly jar fixture mounted over the center of the door. Clearance from the light to the storage space must
be at least 12 inches. Usually, clearance is not an issue with this
method, but if it is, a fluorescent fixture may be used with a minimum clearance of 6 inches. Incandescent fixtures that are recessed
with a completely enclosed lamp may be used with a minimum
clearance of 6 inches to the storage space. Incandescent luminaires
that do not completely enclose the lamp are not permitted inside a
closet. Pendant and open bulb lamp holders are also not permitted
inside a closet.

The control of the closet light can be made by a couple of neat
options that could prove more convenient to some of your customers. Following is a list of possibilities:

- *Use a single pole switch with pilot light. Install it outside the closet on either side of the door. The occupants will know if the closet light has been left on if the pilot light is lit on the wall switch.*

- *Use a single pole plunger switch installed in the door casing on the hinge side. When the door is closed the plunger is held in and the light remains de-energized. When the door is opened, the plunger pops out and power is sent to the light. If the door stays open the occupant will see that the light is on and shut the door.*

- *Larger walk in closets need more light. A 4 foot long, one or two lamp fluorescent wrap around fixture is an excellent choice for years of efficient, maintenance free lighting.*

Attic

When referring to an attic, I'm talking about the area of a house below the roof and above the ceiling of the last floor. This attic is not used for anything. It has open truss work and the floor is covered with insulation. Under normal circumstances nobody ever goes up into this space. If anybody ever goes up there it would probably be to investigate a problem. The problem could be a roof leak, a frozen plumbing vent, a ventilation problem, or birds, rodents, insects, etc. I like to install two or three incandescent lamp holders with 100 watt lamps in this space. Mount them as high as possible, but not higher than would make them hard to change the bulbs from a standing position. The switch should be located near the access hole to the space. Anybody going into the attic can illuminate it before completely entering it. This way, any hazards such as bees nests, bats, or rodents can be seen before it is too late. Another neat idea is to use lamp holders with integral receptacles. This would provide power for additional lighting or power tools that might be needed in the attic.

fastfacts

Permanent light fixtures should normally be placed in the center of the room being served by the light.

Trade Tip: Ceiling lights are not installed nearly as often as they used to be. Some people don't have a problem with switched outlets being used in place of overhead lighting. However, there are still plenty of people who prefer overhead lighting. While you can meet code requirements in many rooms with nothing more than a switched outlet, you should check with your customers to make sure that you offer the option of overhead lighting.

Exterior Lighting

The N.E.C. requires at least one wall switch controlled light at any outdoor entrance at grade level. Of course, we are going to do better than that. First of all, the switch should be located inside next to the strike side of the door. Usually it will be ganged with one or more other switches. The other switches would most likely control foyer, hall, kitchen, or living room lighting. The front door could be flanked by decorative wall lanterns. Sliding glass doors that lead to outside decks and patios will also look elegant when flanked with decorative wall lanterns. Side entrances and doors to utility rooms can be illuminated with a single luminaire mounted on the strike side or centered above the door. In addition, decorative post lanterns can be installed on each side of the driveway and along outside walkways. These lights will add fashion, safely, and security to the property. They can be controlled by a wall switch from inside the home. You could even place a switch outside for the added convenience of controlling the lights from the walkway, garden or patio for example.

Button type photocells are available to install in the light post itself. This way, the lights would go on and off automatically as the sun rises and sets. Another form of outside lighting control is to use an electronic programmable timer. The timer would be preset to turn the lights on at a specific time and off at a specific time. An ideal way to control driveway, walkway, or general yard lighting is to use both a timer and a photocell. The way you do this is to connect the photocell switch in series with the lighting circuit controlled by the timer. Set the timer to come on at 3 p.m. and off at an hour when the occupants would no longer need the lighting. This would typically be around 11 p.m. to 2 a.m., depending upon the habits of the occupants of the home. Now, remember that a photocell has been placed in series with the light circuit. The photocell will not let power flow through itself until it is dark; therefore, the lights will not come on at 3 p.m., but at dark. This control system will end the need to periodically reset the on time in the timer to accommodate for days becoming longer or shorter.

Landscape Lighting

Landscape lighting can dramatically affect the look of a home and property, not to mention the additional safety and security that it provides. There are four basic steps in landscape lighting:

1. What to light?
2. Choosing landscape lighting systems (12 volt or 120 volt)
3. Landscape lighting techniques
4. Landscape lighting plans

Step 1: What to Light?

Examine the home and surroundings. Decide what you would like to illuminate. This could be flowers, trees, shrubs, sculptures, or architectural features of the home. Determine if these features should be highlighted or just provided with general illumination from background lighting. The question to ask your customers is what favorite landscape or architectural features should have the extra brightness of accent lighting.

Step 2: Choosing Landscape Lighting Systems

Before you begin to choose the type of lighting technique, you must decide what voltage of luminaires to use. The two choices are 12

 Trade Tip: Exterior low voltage lighting provides a lot of appeal for a home. It's not uncommon for this lighting option to be overlooked. By making suggestions for low voltage lighting outdoors, you may be able to increase your profit on a job.

volt and 120 volt. Generally speaking, objects 25 feet tall or more and objects over 30 feet away from the light source will be better served with 120 volt systems. Ground mounted 120 volt luminaires must be permanently installed according to local and national codes and manufacturers' instructions. Direct burial cable should be placed one foot or deeper in the earth. Sharp objects and foreign material must be removed from the trench before refilling it. Sections of cable underneath walkways and other traffic areas should be sleeved with PVC conduit. The PVC sleeve should be equipped with an insulating bushing on each end. Splices in the wiring must be waterproof and made by approved methods only.

Systems operating at 12 volts derive their power from a step down transformer that is supplied by a 120 volt circuit. The transformer has a 120 volt primary and a 12 volt secondary. The reduction in voltage also means a reduction in current. There is no risk of injury from electrical shock with the low voltage system. This eliminates the need to bury supply cables as deep. There are drawbacks. Circuits are limited in length and number of luminaires by the transformer used. The advantages are ease of installation and removal. This makes 12 volt lighting a good choice where periodic removal and reinstallation is necessary due to plant growth or heavy snow accumulation.

Step 3: Landscape Lighting Techniques

When lighting pathways, position lights on alternating sides of the path or walk for even illumination. Lights placed along the edge of a driveway should be one foot from the edge. This will define the edge of the driveway for better safety. Lighting for steps must be positioned to avoid shadowing.

Up lights are like footlights in a theater. They focus attention on a specific subject such as walls and trees. Remember, this lighting task will more than likely require 120 volt luminaires; therefore, position lights and wiring to allow for future growth of young plants. Aim fixtures away from the direction of viewers. This technique will help to prevent unwanted glare. Always try to conceal spotlights behind plants to maintain the natural look of the grounds.

Grazing enhances the texture of the brick, stone, or the bark of tree trunks. Fixtures should be placed 6 to 12 inches from the surface. Aim the lights along the surface. Wider beam spreads will illuminate more surface area of a wall or gazebo, for example.

Another technique for highlighting distinctive objects is to silhouette them against a lighted wall. Place luminaires behind plants, shrubs, and small trees that are growing around the perimeter of a building. This technique of lighting the exterior walls also contributes to security lighting.

Step 4: Landscape Lighting Plans

It will help to draw an area layout to scale. Indicate what the homeowner wants to light, what techniques will be used, and if 12 volt or 120 volt fixtures will be used at each location. Remember that a little bit of light goes a long way in the dark. Don't use any more than is necessary at any one location. When using a 12 volt system, you must add total lamp wattage and the distance along the cable from the transformer to the last luminaire on the run. Check this information with the cable/wattage chart to find transformer and cable size required for that circuit.

After the plan is drawn it is easier to select other accessories that are need or desired. Ground stakes, stems, connectors, photocells, and timers would be included in the plan.

RESIDENTIAL LUMINAIRES

Lamp Holders

Ceiling receptacle lamp holders constructed of plastic, fiber, or porcelain, used with incandescent lamps, are the most basic of all luminaires available. They are designed to mount on a 4 inch octagon box. They will also fit on a single gang wall box, which is permissible for wall mounting, as they weigh less than six pounds. Lamp holders are available with built in 15 amp receptacles. Lamp holders without grounded receptacles do not come equipped with a ground terminal because they have no non-current carrying metal parts to ground. If you install a lamp holder with a 15 amp receptacle, the ground terminal must be grounded. In replacement applications where a grounding means is not present, you should use a unit with a two wire non-grounding receptacle built into it.

Common uses for lamp holders would include basements, garages, attics, crawl spaces and generally speaking, any unfinished area where minimal illumination is required. It's important to note that open bulb lamp holders are not permitted in closets.

Jelly Jar Fixtures

A jelly jar light is a simply constructed incandescent lamp holder with a glass guard that totally encloses the lamp. The glass guard is similar in appearance to a jar. These fixtures work very well in closets, cellar or attic stairways, or at exterior entrances. Advantages are low cost and ease of lamp replacement (Figure 7.14).

Close-to-Ceiling Fixtures

These luminaires are excellent for general illumination in any room or area of a house. They are functional and fashionable and come in incandescent and fluorescent styles. Incandescent versions can take one or more lamps. Fluorescent versions are available for use with two and four pin compact lamps and circline lamps. (Figures 7.15a, b, and c).

FIGURE 7.14 Jelly jar type fixtures. *(Courtesy Progress Lighting)*

FIGURE 7.15a Close-to-ceiling fixtures. *(Courtesy Progress Lighting)*

FIGURE 7.15b Close-to-ceiling fixtures. *(Courtesy Progress Lighting)*

FIGURE 7.15c Close-to-ceiling fixtures. *(Courtesy Progress Lighting)*

Pendants

Pendants provide versatile styles of fixtures for dining areas, kitchens, and breakfast nooks. Mini-pendants are ideal for mounting over breakfast bars and kitchen islands. Pendants should be mounted 24 to 30 inches above a tabletop. Mini-pendants should be 18 to 24 inches above a surface area. Pendants can be hung by cord, chain, or rigid stem. (Figure 7.16a through 17.16e.)

FIGURE 7.16a   Pendant style fixtures. *(Courtesy Progress Lighting)*

FIGURE 7.16b Pendant style fixtures. *(Courtesy Progress Lighting)*

FIGURE 7.16c Pendant style fixtures. *(Courtesy Progress Lighting)*

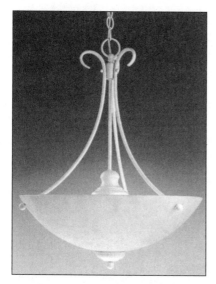

FIGURE 7.16d Pendant style fixtures. *(Courtesy Progress Lighting)*

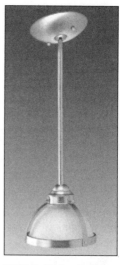

FIGURE 7.16e Pendant style fixtures. *(Courtesy Progress Lighting)*

Chandeliers

Chandeliers come in nearly limitless designs that can be used in many rooms throughout the home. Chandeliers can be used in conjunction with companion fixtures. Companion fixtures are luminaires that are similar in design and style. Examples are wall sconces and close-to-ceiling fixtures. Chandeliers hung over tables should be approximately 6 inches narrower than the smaller dimension of the table. Mount chandeliers 30 inches above a table for best results. (Figure 7.17a, b.)

Wall Lights

Sconces originally were designed to hold torches or candles on the wall in a castle or similar structure. Today, sconces are available in incandescent, halogen, and energy efficient fluorescent versions. Sconces are wall mounted. Another type of wall mounted fixture is the wall bracket. These fixtures are used for general lighting or accent lighting in hallways, stairways, foyers, living and dining rooms. Mount fixtures between 5½ to 6½ feet from the floor to center of the fixture. This will reduce the "cave effect" in corridors often caused by recessed down lights. Accent lighting adds a final touch. Compliment a chandelier with companion sconces or wall brackets (Figures 7.18a, b).

FIGURE 7.17a   Chandelier. *(Courtesy Progress Lighting)*

FIGURE 7.17b   Chandelier. *(Courtesy Progress Lighting)*

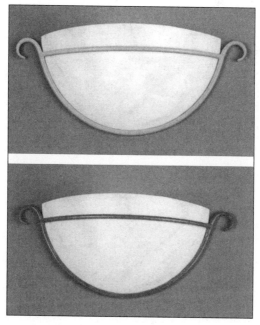

FIGURE 7.18a   Sconces. *(Courtesy Progress Lighting)*

FIGURE 7.18b   Sconces. *(Courtesy Progress Lighting)*

Modular Fluorescents

Energy efficient fluorescents provide up to three times more light than comparable incandescents. In addition, fluorescent lamp life is up to ten times longer than incandescent. New color corrected lamps make these fixtures ideal for the kitchen, bath, laundry, workshop, and family room. Decorative styles are available for more formal areas. Wrap arounds are functional general utility fluorescents designed for individual or continuous row ceiling mounting. They are equipped with white end caps and a clear prismatic acrylic lens (Figures 7.19 through 7.26).

Under-Cabinet Lights

Under-cabinet lights provide supplemental illumination where demanding tasks are performed such as in kitchens and bar areas. Incandescent fixtures are available in 120 volt and 12 volt systems. Fluorescent luminaires are trim and energy efficient. Some models come equipped with a built in switch.

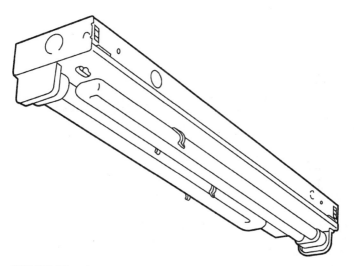

FIGURE 7.19a Compact fluorescent light fixture. *(Courtesy Lithonia Lighting)*

MOUNTING DATA

For unit or row installation, surface or suspended mounting.
Unit installation — Minimum of two hangers required.
Row installation — Two hangers per channel. One per channel plus one per row required if CONLGC installed.
Hooker® (HRC) — Minimum two per channel (unit and row).

DIMENSIONS

Inches (centimeters). Subject to change without notice.

D = 11/16 (1.74) Dia.K.O.
E = 7/8 (2.22) Dia.K.O.
F = 1-1/4 (3.17) Dia.K.O.
H = 2 (5.08) Dia.K.O.

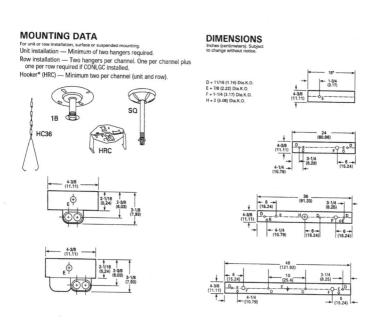

FIGURE 7.19b Compact fluorescent fixture data. *(Courtesy Lithonia Lighting)*

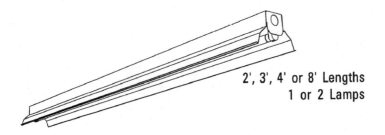

2', 3', 4' or 8' Lengths
1 or 2 Lamps

Specifications

Length: 22-7/16 (569), 34-1/4 (869), 46-1/16 (1169) or 92-1/16 (2337)

Width: 4-3/4 (121), 3-7/16 (87)

Depth: 2-13/16 (71)

Weight: 7.1 lbs (3.2 kg)

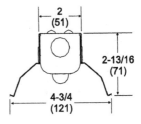

FIGURE 7.20 Low profile fluorescent light fixture. *(Courtesy Lithonia Lighting)*

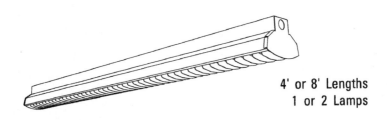

4' or 8' Lengths
1 or 2 Lamps

Specifications

Length: 46-1/16 (1169) or 92-1/16 (2337)

Width: 4-13/16 (122)

Depth: 3-9/16 (90)

Weight: 8.4 lbs (3.8 kg)

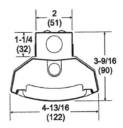

FIGURE 7.21 Low profile louvered fluorescent light fixture. *(Courtesy Lithonia Lighting)*

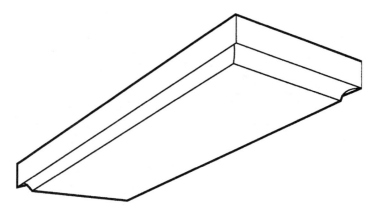

FIGURE 7.22a Steel fluorescent light fixture. *(Courtesy Lithonia Lighting)*

DIMENSIONS

Subject to change without notice.

1' X 2' version overall outside dimensions: 10-1/2" X 24" X 4"
1' X 4' version overall outside dimensions: 10-1/2" X 48" X 4"
1-1/2' X 4' version overall outside dimensions: 16" X 48" X 4"
2' X 2' version overall outside dimensions: 24" X 24" X 4"

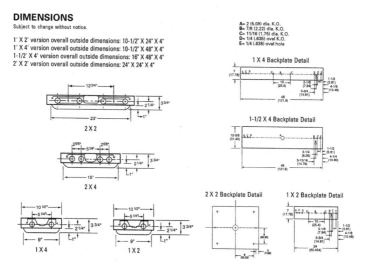

FIGURE 7.22b Steel fluorescent fixture data. *(Courtesy Lithonia Lighting)*

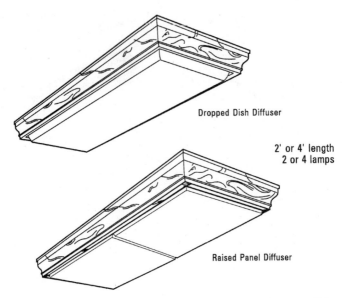

Dropped Dish Diffuser

2' or 4' length
2 or 4 lamps

Raised Panel Diffuser

FIGURE 7.23a Wood framed fluorescent light fixture. *(Courtesy Lithonia Lighting)*

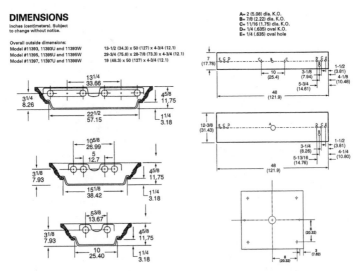

FIGURE 7.23b Wood framed fluorescent light fixture data. *(Courtesy Lithonia Lighting)*

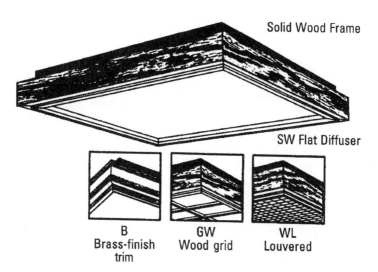

FIGURE 7.24a Wood frame fluorescent fixture options. *(Courtesy Lithonia Lighting)*

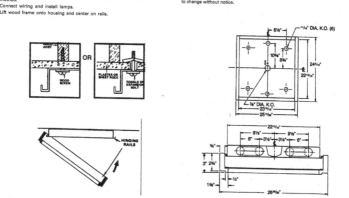

MOUNTING DATA

1. Secure housing to ceiling with wood screws, toggle bolts or similar method.
2. Connect wiring and install lamps.
3. Lift wood frame onto housing and center on rails.

DIMENSIONS

Inches (centimeters). Subject to change without notice.

FIGURE 7.24b Wood frame fluorescent fixture mounting data. *(Courtesy Lithonia Lighting)*

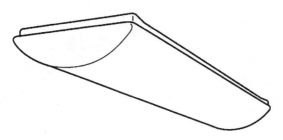

FIGURE 7.25a Low profile surface mount fluorescent light fixture. *(Courtesy Lithonia Lighting)*

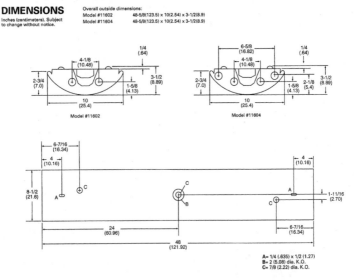

DIMENSIONS

Inches (centimeters). Subject to change without notice.

Overall outside dimensions:
Model #11602   48-5/8(123.5) x 10(2.54) x 3-1/2(8.9)
Model #11604   48-5/8(123.5) x 10(2.54) x 3-1/2(8.9)

Model #11602

Model #11604

A= 1/4 (.635) x 1/2 (1.27)
B= 2 (5.08) dia. K.O.
C= 7/8 (2.22) dia. K.O.

FIGURE 7.25b Low profile surface mount light fixture dimensions. *(Courtesy Lithonia Lighting)*

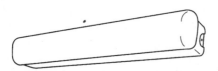

FIGURE 7.26a Wall mounted fluorescent light fixture. *(Courtesy Lithonia Lighting)*

DIMENSIONS

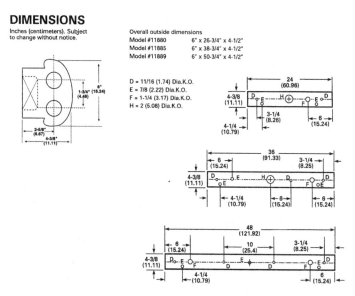

FIGURE 7.26b   Wall mounted fluorescent light fixture dimensions. *(Courtesy Lithonia Lighting)*

Exterior Luminaires

Fixtures used outdoors must be labeled for the use intended. They are designed to be weather tight. Incandescent fixtures come in modest to extremely ornate designs. Exterior fixtures are also available in energy efficient compact fluorescent. These fixtures offer excellent performance, low operating cost, and average lamp life of 10,000 hours. This feature makes them an ideal choice for installing in locations where lamp replacement is a major task (Figure 7.27).

Landscape Lighting

Landscape lighting comes in 120 volt and 12 volt systems. Generally speaking, 120 volt luminaires are used to light larger objects or objects that are far away. 12 volt systems are safer to use around heavy root growth because the wires don't have to be buried as deeply. Landscape lighting is used to light walkways, trees, plant growth, statues, fountains and architectural features of buildings (Figures 7.28 through 7.32).

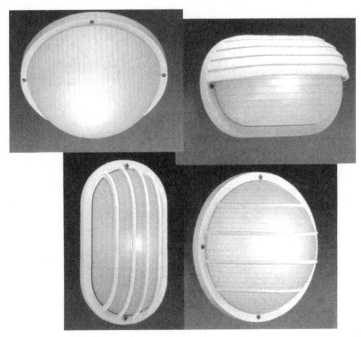

FIGURE 7.27 Exterior light fixtures. *(Courtesy Progress Lighting)*

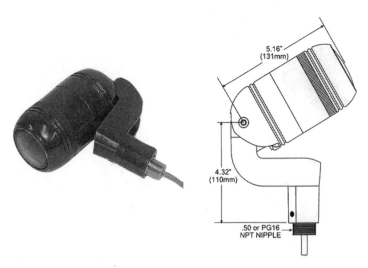

FIGURE 7.28 Twelve volt exterior accent light. *(Courtesy Lithonia Lighting)*

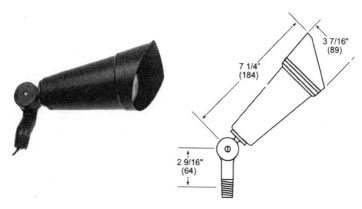

FIGURE 7.29 Exterior accent light. *(Courtesy Lithonia Lighting)*

SPECIFICATIONS:

MATERIAL: Extruded aluminum with die cast aluminum end caps.
Knuckles cast bronze.

LAMP: T-10, 60 Watt Max. (IN);
E-17 (HP);
13TT (Compact Fluorescent); **Lamp by others.**
T-8, 32 Watt (FL);
T-12, 40 Watt (FL).

VOLTAGE: 120V.

SOCKET: Pulse-rated medium base (IN, HPS);
GX23D (Compact Fluorescent);
Medium Bi-pin (FL).

FINISH: Bronze (BZ) textured powder coat standard. Other colors
available.

LIGHT DISTRIBUTION: Flat stainless steel reflector - Flood.

LENS: Tempered glass (IN, HPS).
Clear acrylic (Compact Fluorescent, FL).

MOUNTING: 4798 - One 1/2" NPT adjustable cast bronze knuckle.
4799 - Two 1/2" NPT adjustable cast bronze knuckles.

15° Vertical Plane
Fixture Lens

U.L. LISTED
FOR WET LOCATIONS

DIMENSIONS:

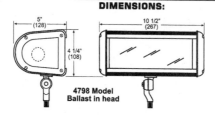

5"
(128)
10 1/2"
(267)
4 1/4"
(108)
**4798 Model
Ballast in head**

FIGURE 7.30 Exterior accent light data and dimensions. *(Courtesy Lithonia Lighting)*

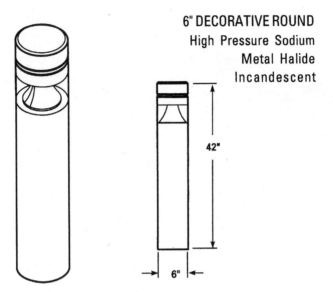

6" DECORATIVE ROUND
High Pressure Sodium
Metal Halide
Incandescent

42"

6"

FIGURE 7.31a Round exterior architectural bollard light fixture. *(Courtesy Lithonia Lighting)*

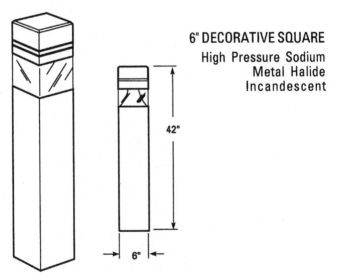

6" DECORATIVE SQUARE
High Pressure Sodium
Metal Halide
Incandescent

42"

6"

FIGURE 7.31b Square exterior architectural bollard light fixture. *(Courtesy Lithonia Lighting)*

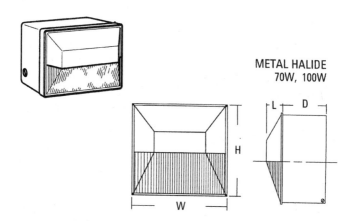

METAL HALIDE
70W, 100W

Standard Dimensions

Height: 12-1/4 (31.1)
Width: 14-1/4 (36.2)
Depth: 8-3/16 (20.8)
Back Box: 6 (15.2)
Lens: 2-3/16 (5.6)
Weight: 15 lbs. (7 kg)

All dimensions are inches (centimeters) unless otherwise specified.

Bi-adjustable socket assembly
KL "D" Series

FIGURE 7.32a Recessed low mount exterior light fixture. *(Courtesy Lithonia Lighting)*

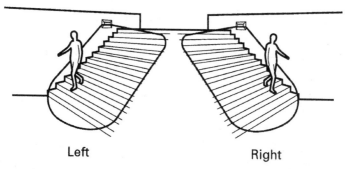

Left

Right

FIGURE 7.32b Directional light distribution from recessed low mount exterior light fixture. *(Courtesy Lithonia Lighting)*

Track Lighting

The track is a metal channel with isolated conductors inside it. It can be custom fit for the area it will be used in. For example, it can be run around the perimeter of a living room. Turns and tees are made with special fittings designed for use with the track. You can swivel, rotate, and aim lamp holders in any direction for general illumination, wall washing, or accent lighting. Halogen, fluorescent, and incandescent models are available (Figures 7.33 through 7.35).

Recessed Lighting

Recessed fixtures are very popular today. Versatility and low ceiling profile are the main reasons for this popularity. There are three basic components of this class luminaire (Figures 7.36 through 7.38):

FIGURE 7.33a   Track light fixture. *(Courtesy Progress Lighting)*

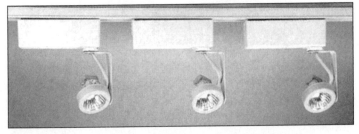

FIGURE 7.33b   Track light fixture. *(Courtesy Progress Lighting)*

1. *Housing:* this is the main part of the fixture. It consists of the frame, protective shell, junction box, and lamp holder. Many housings are universal and will accept a multitude of trims. Housings are available for new construction. They must be installed before the finished ceiling is in place. Housings for remodeling are designed to be cut in and supported by the finished ceiling. Units are available for direct contact with insulation and combustible material.

2. *Trim:* this is the finished part that is visible on the ceiling. Trims range from a basic cone or step baffle to directional eyeballs and shower leases. Each trim will include a list of acceptable lamp types and wattages.

3. *Lamp:* depending on the listings of the housing and trim used, several lamp styles can be used with each luminaire. Examples of commonly used lamps are A-19, PAR-20, R-20, PAR-30, BR-30, PAR-38, and BR-40. Recessed housings and trims are also designed to use compact fluorescent lamps. Models are also available for use in sloped ceilings.

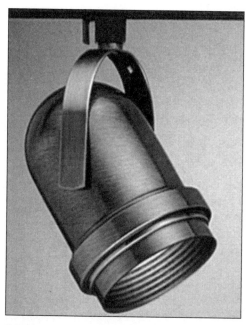

FIGURE 7.34   Track light swiveling detail. *(Courtesy Progress Lighting)*

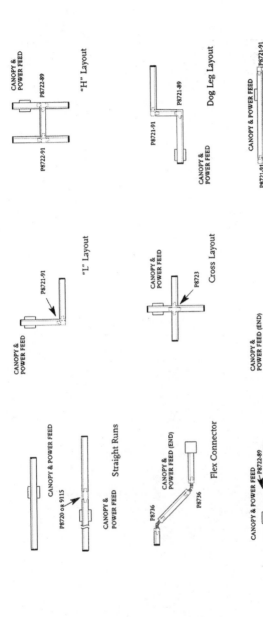

CANOPY &
POWER FEED
P8722-89

P8722-91

"H" Layout

P8721-91

"L" Layout

CANOPY &
POWER FEED

CANOPY &
POWER FEED

P8720 or 9115

Straight Runs

CANOPY &
POWER FEED

P8721-89

Dog Leg Layout

CANOPY &
POWER FEED

CANOPY &
POWER FEED

P8723

Cross Layout

CANOPY &
POWER FEED (END)

P8736

P8736

Flex Connector

CANOPY & POWER FEED

P8721-91   P8722-89   P8721-91

DEAD END

P8721-91

P8721-91

P8721-91

Modified Grid Layout

CANOPY &
POWER FEED (END)

P8722-91

"T" Layout

P8721-91

CANOPY & POWER FEED

P8722-89   P8721-91   P8722-91   P8721-91

P8723

P8722-91

P8721-91   P8722-89   P8721-91

Grid Layout

FIGURE 7.35 Track lighting layout. *(Courtesy Progress Lighting)*

FIGURE 7.36a   Recessed lighting enclosure rough-in. *(Courtesy Progress Lighting)*

FIGURE 7.36b   Recessed lighting enclosure. *(Courtesy Progress Lighting)*

FIGURE 7.37 Self-contained recessed lighting enclosures save labor and extra materials. *(Courtesy Progress Lighting)*

CEILING PADDLE FANS

Ceiling fans can be used with integral light fixtures attached. Units are available with nightlights as well. Fan/light units are inexpensive and can be very ornate. They can be wired with speed switches to control the fan as well as dimmer switches to control the level of light. Wireless remote control components can be installed in most fans. Ceiling fixtures give good general illumination and the air movement generated by the fan keeps the room comfortable year around.

BATHROOM EXHAUST FANS

Fan units are available with combination light, night light, and electric unit heater. The combination fan/light is an excellent fixture to

FIGURE 7.38 Recessed lighting installed in sloped ceiling. *(Courtesy Progress Lighting)*

mount in the center of a bathroom. Moisture and odors are carried out of the room and blown outside through a simple flexible duct system. The centrally located light provides general illumination for most tasks in the bathroom. It is recommended to have a separate switch for each feature of a combination unit.

EMERGENCY LIGHTING

Emergency units can be purchased that are compact and attractive. Units are available for wall or ceiling mounting. Recessed emergency lights can be installed in a ceiling. They have a low profile gimbal ring for a finish trim. We recommend emergency light units in bathrooms, stairways, and in the general area of the main electrical panel. Just think how convenient this will be if the homeowner is in the tub or shower when the power goes out. The bathroom would be illuminated enough to get out and dry off safely. The homeowner then can check the circuit breakers to see if the outage is caused by a tripped circuit breaker.

COMMERCIAL LIGHTING

Commercial lighting is more often accomplished with fluorescent and HID luminaires. Many commercial buildings have suspended acoustical ceilings. This makes the job a breeze for the electrician. Removable ceiling tiles allow you access at nearly any location in the room. Please note, however, that above the ceiling are mass quantities of HVAC ducts, plumbing and heating piping, and sprinkler piping. The lighting system circuitry and junction boxes must be laid out so as not to interfere with the other systems. Junction boxes, especially, must be accessible for future inspection and maintenance. The 2002 N.E.C. prohibits the use of nonmetallic sheathed cables above suspended ceilings in other than dwellings. This leaves you with the options of metal sheathed cable such as AC or MC, or conduit.

Power supplies for commercial buildings are commonly 120/208 volt 3 phase 4 wire and 277/480 volt 3 phase 4 wire. Lighting circuits derived from either system consist of one hot phase conductor and the neutral conductor.

Fluorescent ballasts produce harmonic currents which do not cancel each other like resistive loads. You should always run a separate neutral conductor with each phase conductor. Service and feeder cables carrying power for large fluorescent loads must have appropriately sized neutral conductors. Cables are available with super sized neutral conductors for this purpose. The most commonly used luminaires for these applications are fluorescent troffers. These fixtures come in a variety of sizes such as 1 x 4, 2 x 2, 2 x 4, and 4 x 4. The grid troffer fits into the ceiling framework or grid, as it is called. These fixtures are available with numerous options. The option most popular with electricians in the field is integral T bar safety clips. This meets the code requirement for securely attaching the fixture to the grid work. Other options are:

- *Number of lamps*
- *Lamp type*
- *Diffuser type*
- *Frame type*

- *Voltage*
- *Door frames*
- *Reflective surfaces*

Another feature popular in the field is that these fixtures permit in and out through wiring. This greatly reduces the number of junction boxes needed for the job.

Office Lighting

The greatest challenge in lighting today's modern work place is to create an environment that is safe, comfortable and maximizes worker productivity. This task becomes increasingly more difficult as the number of video display terminals in the workplace continues to rise, because of the problems caused by reflections from overhead lighting on VDT screens. Reflected glare on a computer screen can blur text or graphics, increase the number of errors, and cause a decrease in worker productivity. More importantly, reflected screen glare is a proven cause of eyestrain, irritability, fatigue, and longer term health problems.

Traditional lighting systems may no longer provide appropriate solutions because they cannot properly control reflected glare. With the unique Optimax Light Control System from Lithonia Lighting, designers finally have a solution to this lighting challenge.

Optimax is a fluorescent lighting system that eliminates objectionable VDT screen glare caused by luminaire reflections. The key to Optimax performance is the combination of an optimum shielding design that controls light at glare producing angles and precise optical assemblies made specially formulated, low iridescent, highly specular anodized aluminum. As a result, Optimax encourages productivity by eliminating the reflected luminaire screen glare that workers find uncomfortable and irritating (Figures 7.39a, b). Optimax is a unique recessed fluorescent lighting system that meets the three basic lighting requirements of the electronic office:

1. Effectively eliminates objectionable glare on VDT screens caused by reflection from ceiling lights.
2. Delivers appropriate levels of general illumination for non VDT office tasks.
3. Provides economical, energy efficient system performance.

Commercial Surface Mounted Fluorescents

Architectural wraparounds are designed for surface mounting or they can be suspended by stem mounting. They can be installed as single units or in continuous rows of two or more. Corridors are usually illuminated with one or two lamp fixtures, while office or work areas are usually illuminated with two, three, or four lamp fixtures. Continuous row installations are convenient to wire because circuit conductors can be run through the fixtures. This means that only one ceiling outlet needs to be installed per row (Figures 7.40, 7.41a, b).

Striplights are basic fluorescent luminaires. They consist of a metal channel with a baked enamel finish. Units are available in 1, 2, 3, and 4 lamp versions. Tubes are held in place by lamp holders in each end. Tubes are completely exposed. Reflectors and wire guards can be used with strip lights where lamps could be subject to damage.

Wet location fluorescent fixtures have a fiberglass reinforced polyester housing and acrylic diffuser. The diffuser is secured to a fully gasketed housing. These fixtures are designed for surface, stem, or chain mounting. They are most commonly used in areas of high water vapor and where hosing is done.

There are many other styles of commercial and industrial fluorescent luminaires. Consult a catalog from a manufacturer such as Lithonia Lighting to find fixtures, accessories and options that are available for any application.

High Intensity Discharge (HID) lighting is used extensively in commercial and industrial applications. Recessed down lights are popular in malls and other retail stores. Many are UL listed for through branch wiring and damp locations. This makes them an excellent choice for exterior soffit lighting.

Metal halide and high-pressure sodium (HPS) styles are available in a wide range of lamp wattages. Arm mounted area lights are used

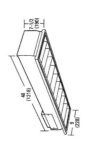

LITHONIA LIGHTING®

Optimax® Parabolic Light Control System

PMO 9"x4'

Symmetric Distribution

FEATURES & SPECIFICATIONS

INTENDED USE

Light controlling parabolic luminaires designed to control screen glare in VDT open office environments. Ideal continuous row systems for general illumination.

ATTRIBUTES

Models available to meet RP-1 minimum and preferred luminance criteria for offices. Efficiently delivers appropriate illumination level for paper-based tasks. Choice of diffuse or specular louvers utilizing the latest developments in louver finishing for minimized louver iridescence. Ideal for use with triphosphor lamps.

CONSTRUCTION

Black reveal provides floating louver appearance, conceals optional air-supply slots. Air flow control provided with optional heat removal dampers and air-pattern control blades.

Overlapping flange and modular ceiling trims factory installed with standard swing-gate hangers or field convertible with optional trim and hanger kits.

T-hinges die-formed for maximum strength. Latches spring-loaded, concealed in the reveal. Ballast boxes used to encase the ballasts, mounted to the top (standard) or side of housing.

Housing formed from cold-rolled steel. Louvers formed from anodized aluminum. No asbestos is used in this product.

FINISH

Five-stage iron-phosphate pretreatment ensures superior paint adhesion and rust resistance. Painted parts finished with high-gloss, baked white enamel.

ELECTRICAL SYSTEM

Thermally-protected, resetting, Class P, HPF, non-PCB, UL Listed, CSA Certified ballast is standard. Energy saving and electronic ballasts are sound rated A.

Luminaire is suitable for damp locations. AWM, TFN or THHN wire used throughout, rated for required temperatures.

LISTING

UL Listed and CSA Certified (see Options). NOM Certified for Mexico (see Options).

WARRANTY

Guaranteed for one year against mechanical defects in manufacture.
Specifications subject to change without notice.

Specifications
Length: 48 (1218)
Width: 9 (228)
Depth: 7-1/2 (190)
Weight: 22 lbs (10 kg)

48
(1218)

9
(228)

7-1/2
(190)

All dimensions are inches (millimeters).

FIGURE 7.39a Optimax parabolic light control system. (*Courtesy Lithonia Lighting*)

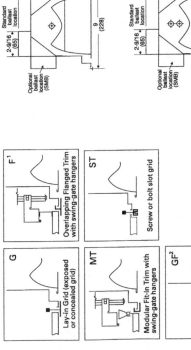

MOUNTING DATA

Continuous row mounting of flanged units requires CRE and CRM options (See Options).

G — Lay-in Grid (exposed or concealed grid)

MT — Modular Fit-In Trim with swing-gate hangers

GF² — Overlapping Flanged Trim (Sides only) to support ceiling tile

F¹ — Overlapping Flanged Trim with swing-gate hangers

ST — Screw or bolt slot grid

DIMENSIONS

7-1/2 (190)
6-3/16 (157)
2-9/16 (65)
9 (228)
Standard ballast location
Optional ballast location (SMB)

7-1/2 (190)
6-3/16 (157)
2-9/16 (65)
9 (228)
Standard ballast location
Optional ballast location (SMB)

NOTES:

1 Recommended rough-in dimensions for F trim fixtures 9"×48" (Tolerance is +1/4", -0"). Swing-gate range 1" to 3-1/4", span 8-1/2" to 11-5/16".

2 GF1 trim option provides flange trim on one side only.

FIGURE 7.39b Optimax mounting data and dimensions. *(Courtesy Lithonia Lighting)*

142

to illuminate streets, walkways, parking lots, and car lots. Where color is not a concern, HPS luminaires provide the best efficiency. Economy of operation and extended lamp life are what make HPS lighting so popular. Where white light is a must, metal halide would be the better choice.

Architectural flood lighting is used for landscape and facade lighting. General purpose flood lights are used to light industrial and commercial yards, construction sites, parking lots and recreational areas. Wall mounted HID fixtures are used for perimeter security lighting, and to light loading docks and vehicle ramps. Wire guards and vandal guards are options available for use with HID fixtures that are subject to physical abuse.

High bay reflector fixtures are used in areas with high mounting heights. These highly efficient luminaires are ideally suited for use in heavy manufacturing areas, retail and warehouse aisles, and gymnasiums. Low bay reflector fixtures are designed to be used in areas with lower mounting heights. A multitude of options is available to meet any commercial or industrial need.

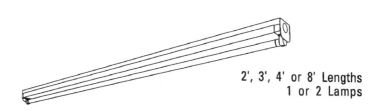

2', 3', 4' or 8' Lengths
1 or 2 Lamps

Specifications

Length: 22-7/16 (569), 34-1/4 (869), 46-1/16 (1169) or 92-1/16 (2337)

Width: 2 (51)

Depth: 2-1/4 (57)

Weight: 4.8 lbs (2.2 kg)

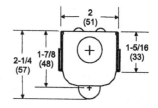

FIGURE 7.40 Low profile direct channel strip fixture. *(Courtesy Lithonia Lighting)*

Rapid Start

2', 3' or 4' length
1 or 2 lamps

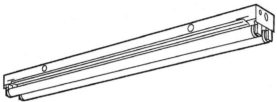

ENERGY

- Luminaire Efficacy Rating (LER)

  One-lamp LER.FS = 52. Two-lamp LER.FS = 59.
  Based on 34W T12 lamp, 2650 lumens, and energy-saving magnetic ballast.
  Ballast factor = .87. Input watts = 42 (one lamp), 71 (two lamps).

  Calculated in accordance with NEMA standard LE-5. See LER sheet for details.

FIGURE 7.41a General purpose rapid start fluorescent strip fixture. *(Courtesy Lithonia Lighting)*

Emergency Lighting

Codes require that means of egress must be illuminated in the event of a power outage. Exits must also be illuminated and clearly identified by a sign. Exit signs come in single and double face styles. They can be top mounted or end mounted. Single face units can also be back mounted. Exit signs can also be recessed. LED type exit signs can have a rated lamp life of 25 years or more. Exit signs with built in emergency lamp heads make the electrician's job easier.

Emergency lighting units are either hardwired or attached by cord and plug to the lighting circuit for the area they will illuminate. When the power to the light circuit goes out, the emergency battery units automatically come on. That is why they must be connected to the lighting circuit ahead of any switching. Codes require that emergency illumination shall be arranged so that the failure of any one head will not leave an area in total darkness. That is why self contained units have two or more heads.

MOUNTING DATA

For unit or row installation, surface or suspended mounting.

Unit installation — Minimum of two hangers required.

Row installation — Two hangers per channel required. One per fixture plus one per row if CONLGC installed.

Hooker® (HRC) and HC Hangers — Minimum two per channel (unit and row)

See ACCESSORIES below for hanging devices.

DIMENSIONS

Inches (centimeters). Subject to change without notice.

Aluminum channels have different mounting details:
48", 72" and 96" have only two 7/8" K.O.'s 6" from each end
24" and 36" have only two 7/8" K.O.'s 3-1/4" from each end

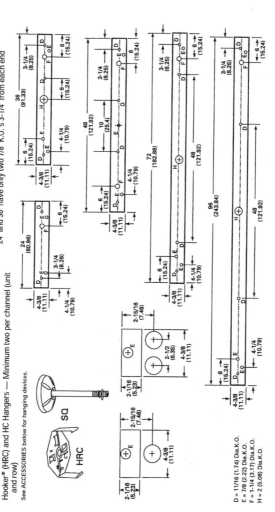

D = 11/16 (1.74) Dia.K.O.
E = 7/8 (2.22) Dia.K.O.
F = 1-1/4 (3.17) Dia.K.O.
H = 2 (5.08) Dia.K.O.

FIGURE 7.41b General purpose rapid start fixture data. (*Courtesy Lithonia Lighting*)

145

Larger buildings with longer means of egress often use large remote battery units and remote heads. In this application, an appropriately sized battery unit is located in a mechanical room or utility closet. Remote heads are mounted in the required areas and wired to the battery unit. Remote heads can be single or double. Remote fixtures also come in ceiling recessed styles. Conductors supplying remote heads are generally larger because they carry low voltage and high current.

Another way to meet code requirements for emergency illumination is to use fluorescent battery packs. Units are available to supply emergency power for:

- *Four foot T8 lamps*
- *Quad tube compact fluorescent lamps*
- *Twin tube compact fluorescent lamps*
- *Triple tube compact fluorescent lamps*

Battery packs look similar to a ballast and will supply power to their lamps for a minimum of ninety minutes. They can be mounted externally or inside the fixture housing.

Emergency lighting in commercial and industrial buildings is very important for safety in the event of an evacuation during a normal power failure. As an electrician, plan the emergency lighting system to be convenient, safe, efficient, and with the capability to allow for future expansion. Never install the bare minimum requirements.

fastfacts

One of the keys to a successful job, whether residential or commercial, is a strong working plan. Many experienced electricians don't bother to draw up a wiring diagram for a simple wiring job. They are likely to wire an entire house without a wiring plan. While the electricians are capable of wiring a house without a diagram, it doesn't mean it is wise. You can normally do a job better, more efficiently, and less expensively when you have a defined plan to work from. It pays to sit down the night before starting a new job and draw out your circuits and homeruns on paper. When you get to work in the morning, there is no guesswork involved.

chapter 8

ELECTRIC HEATING UNITS

lectric heating units are very popular in warm climates where minimal heat is needed during a heating season. The low installation cost of electric baseboard heat is a major reason for this popularity. However, electric heat can be very expensive to use in cold climates. Even so, you can find electric heat in most any state.

The two types of electric heat normally installed are baseboard heating units and wall-mounted heaters. The wall-mounted heaters are quite common in bathrooms. Wall-mounted heaters that are equipped with fans can produce a lot of warmth in a short period of time. This is ideal when taking a bath or shower. Electric heating units are available as built-in features of bathroom exhaust fans. Combination units with exhaust fan, light, night light and electric heater are very popular today.

Electric heat is also popular in remodeling jobs. It is not always possible or practical to tap into an existing heating system to provide heat for a new addition, an attic conversion, or a basement conversion. When this is the case, electric baseboard heat is an economical solution to a potentially expensive heating problem.

If you are thinking of installing electric heat in an existing building, you must check the panel box to see if it has sufficient capability for handling the heavy amps that electric heat will be pulling. You might have to upgrade the electrical service to accommodate the needs of electric heat. Don't make the mistake of bidding a job where electric heat is to be installed without being sure that the existing electrical service and panel box can handle the improvement. You will find such a mistake to be costly.

BASEBOARD HEATING UNITS

Baseboard heating units come in lengths that range from 2 to 10 feet. They are usually 7 or 8 inches high and about 2 to 3 inches in depth. Most units require 240 volts of electricity and pull about 1 amp per linear foot of baseboard. When sizing a circuit, don't forget to allow for the safety factor of loading a circuit to no more than 80 percent of its rated ability. In other words, a 20 amp circuit should not be subjected to more than 16 amps. This means that you could install approximately 16 linear feet on a 20 amp circuit.

Be careful when selecting your baseboard units. Some units are made to work on 120 volt systems. Don't make the mistake of installing a 120 volt section of baseboard heat on a 240 volt circuit. If you do, the heating unit may overheat and could start a fire. Very few people install 120 volt units, since they pull a lot of power and tend to be inefficient.

Baseboard heat is normally wired in parallel. Twelve gauge cable is normally used to wire the heating units. Preferred placement of the heating units is on outside walls, below windows. Avoid placing baseboard heat under electrical outlets. Cords plugged into an outlet could come into contact with the heating unit and create a hazard. You might have to install shorter sections on either side of an outlet to avoid unnecessary risk with cords that may be plugged into outlets.

fastfacts

Large fuses, such as those used with 240 volt circuits, can be protected by pullout blocks that contain cartridge fuses. When a pullout block is not present and you have to pull a cartridge fuse, don't use your fingers. A fuse puller should be used to ensure your safety. Always make sure the fused disconnect handle is in the off position.

fastfacts

What gauge wire is normally used with 240 volt systems? Number 6 wire can handle 240 volts and up to 60 amps. A number 8 wire works with 240 volts and up to 40 amps. Number 10 wire is good to go with 240 volts and up to 30 amps. Generally speaking, all building wiring is rated for up to 600 volts. The gauge of the wire is determined by the current draw in amperes. A water pump rated at 240 volts and drawing 5 amps can be supplied by a 14 AWG conductor.

Thermostats for electric heaters are normally mounted on the heating units. They could be mounted on a wall, but they rarely are. Double-pole thermostats are safer than single-pole thermostats. The safety factor is not so much about safety in the use of the heater. A single-pole thermostat installed on a 240 volt heater maintains power at the heater even when the heater is not on. Anyone working on the heater might think that the power is off and wind up with a serious shock. Always cut the power off to heating units at the service panel before working on the units.

Wall Heaters

Wall heaters are very handy when heating a bathroom or other small area. Most wall heaters have fans and thermostats. Heaters with a 250 volt rating use less electricity than those with a 120 volt rating. Also, never install a 120 volt heater on a 240 volt circuit.

Typically, wall heaters are mounted in a metal box. The metal box is roughed into the framed wall cavity before drywall is hung. Wall heaters range in size from 750 watts to 1,500 watts. If you figure 1 amp per 250 watts, you can see that even a 1,500 watt heater only pulls 6 amps. As long as the total amp load doesn't exceed 16 amps, more than one heater can be put on the same circuit. When wired on its own circuit, a wall heater should be wired with a 12 gauge cable.

Depending upon where you work, you may not see a lot of demand for electric heat, but you will probably work with it occasionally. The wiring process is not complicated, but you do have to make sure to follow the rules that we have discussed here.

chapter

9

HVAC ELECTRICITY AND WIRING

VAC systems use high-voltage systems for running motors, compressors and other big equipment. These devices operate at voltages anywhere from 120 V up to 480 V and even higher. On the other end of the scale, many electronic control circuits use lower voltages in the 12 to 24 V range for running actuators, sensors, and controllers. This chapter discusses some of the basics of electronic circuits and equipment.

Electricity is often compared to water. While working with the two together can be a dangerous combination, the analogy between the two is quite workable. Just as water flows through a pipe, you can think of electrons flowing through wires. Water current is measured in gallons per minute, while current flow is measured in electrons per second. Water pressure is measured in pounds per square inch, while electricity pressure is measured in volts. Table 9.1 gives some equivalents that may make is easier to understand the various electrical terms. Of course, the analogy can only be taken so far. In the case of alternating current, it is difficult to imagine a pipe in which the pressure changes from positive to negative many times per second.

TABLE 9.1 Equivalents for Electricity

Term	Description	Units
energy	joules or Btus	1 Btu = 1055 joules = 0.293 Wh
coulomb	number of electrons	1 coulomb = $6 \cdot 10^{23}$ electrons
current	flow rate of electricity	1 ampere (1 A) = 1 coulomb / second
volts	"pressure" of electricity	1 volt (1 V) = 1 joule / coulomb
resistance	obstruction of current	1 ohm (1 Ω) = 1 volt / 1 ampere
capacitance	"pressurized storage" of electricity	1 farad (1 F) = 1 coulomb / volt
power	rate of energy	1 watt (1 W) = 1 joule / second

There are many different types of electronic components. Most modern electronic HVAC controllers use a combination of *semiconductors* to process and generator electronic signals. The operation of these devices can be pretty complex, so we're not going to cover them in this book. However, there may be times when you need to identify certain components on a wiring diagram so that you can track down problems. Table 9.2 gives a description of different electrical components that you may encounter.

DC CIRCUITS

Direct current (DC) circuits are used in HVAC systems primarily for sensors, electronic controllers, and electronic actuators. The basic operation of DC circuits is governed by two concepts: (1) Ohm's law, which states that the voltage drop across a resistor found by multiplying the current i and the resistance R, $V = i \times R$, and (2) Kirchhoff's voltage law, which says that the sum of voltages around a closed loop must be equal to zero. Using the water analogy, these two laws say that there is a electrical pressure drop when the current passes through a valve, and that the total pressure rises around a circuit must add up to all the pressure drops. Resistors are usually color-coded (see Figure 9.1) to identify both the magnitude of the resistance as well as the accuracy of the rated resistance.

TABLE 9.2   Electronic Component Symbols

Component	Symbol	Function
Capacitor		Maintains constant voltage when current varies
Circuit breaker		Breaks current flow when current is too high
Diode		Limits current flow to one direction
Fuse		Breaks current flow when temperature gets too high
Ground		Reference for voltage in circuit
Inductor		Maintains constant current when voltage varies
Meter		Measures voltage, current, or total power consumption
Operational amplifier		Amplifies an electronic signal
Relay		Switches current between two or more different terminals
Resistor		Impedes the flow of electricity
Switch		Makes or breaks the continuity of a circuit
Transformer		Converts AC power at one voltage to another voltage
Transistor		Electronic switch/amplifier
Voltage source		Provides voltage to a circuit

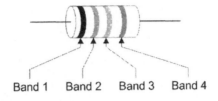

Color	First band	Second band	Third band	Fourth band
	First digit	Second digit	Multiplier	Tolerance
none				± 20%
Silver			0.01	± 10%
Gold			0.1	± 5%
Black	0	0	1	
Brown	1	1	10	± 1%
Red	2	2	100	± 2%
Orange	3	3	1,000	
Yellow	4	4	10,000	
Green	5	5	100,000	± 0.50%
Blue	6	6	1,000,000	± 0.25%
Violet	7	7	10,000,000	± 0.10%
Grey	8	8		± 0.05%
White	9	9		

FIGURE 9.1   Resistor showing location and value of color bands.

Some common resistor color combinations are:

- Yellow/Violet/Black = 47Ω
- Brown/Black/Brown = 100Ω
- Yellow/Violet/Brown = 470Ω
- Green/Black/Brown = 500Ω
- Brown/Black/Red = 1000Ω
- Green/Black/Red = 5000Ω
- Brown/Black/Orange = 10000Ω
- Green/Black/Orange = 50000Ω

Many temperature sensors used in HVAC measurements are resistors where the resistance changes according to the temperature. Remember that resistors act like a valve in a water stream—they decrease the voltage. Resistors behave differently when connected in series or in parallel (see Figure 9.2). The total resistance of series resistors is simply the sum of the individual resistances, $R_{TOTAL} = R_1 + R_2$, where the total resistance of the parallel resistors is taken from the sum of the reciprocals, $R_{TOTAL} = (1/R_1 + 1/R_2)^{-1}$

Voltage dividers are made up of two series that reduce the voltage provided to a sensor or controller. For DC circuits, a voltage divider works as illustrated in Figure 9.3. In some HVAC control loops you find slide wire resistors (Figure 9.4). These are resistors with three connections: one at each end and a movable tap in the middle. The resistance between the terminal at either end and the tap depends on the tap position. Standard end-to-end resistances of slide wire resistors are 100Ω, 135Ω, and 1000Ω.

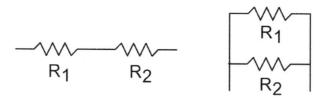

FIGURE 9.2   Series (left) and parallel (right) resistors.

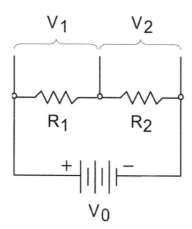

FIGURE 9.3   DC voltage divider.

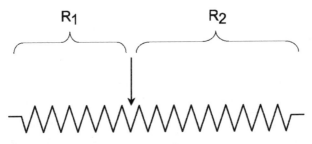

FIGURE 9.4 Slide wire resistor.

Suppose you had a controller that needed to know the difference between the outside air temperature and the supply air temperature in a building. You could use a Wheatstone bridge like that shown in Figure 9.5. These kinds of devices are often used in HVAC sensors that must respond to a small change in voltage or current. If all the resistances in this circuit are equal, then the output voltage V_o equals zero. Now suppose R3 is a variable resistor that measures the outside air temperature (this is called the compensation resistor, R_c) and R2 is a variable resistor that measures the supply air temperature. The output voltage of the bridge varies depending on the difference in resistance.

Capacitors are used to maintain a constant voltage under varying current conditions. They are an integral part of electronic filters that can are used, for example, to eliminate noise from sensor measurements using a filter like in Figure 9.6. This is called a low pass filter because low frequency signals are not affected while high frequency signals (the noise) get eliminated.

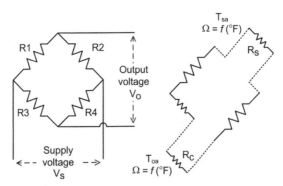

FIGURE 9.5 Schematic of a Wheatstone bridge circuit.

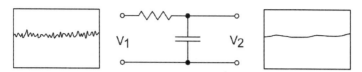

FIGURE 9.6 Low-pass filter.

A diode is like a check valve in that it only allows current to flow one way but not the other. You can use diodes to create a DC power supply from an AC line signal to use for powering sensors. A *rectifier* (Figure 9.7) is used to do this. The AC signal enters at the left and passes through the diodes that are used to create a full-rectified signal. This is then passed through a low-pass filter as described in the previous paragraph.

A transistor is like a valve with three connections: the collector, the emitter, and the base. When a current is applied to the base, the valve opens up and lets current flow from the collector to the emitter. In fact, even the current applied at the base goes out the emitter. In some DC circuits, power transistors are used to amplify sensor or control signals to useful levels. These components can produce a lot of heat that must be dissipated to prevent burning up the transistor. Often there is a heat sink (see Figure 9.8) attached to the side of a control box. Always insure that the heat sinks are clean and in a free air stream. Do not cover them with rags, papers, or anything else. If the heat sink is not hot, then it's doing its job. By covering it up, you are risking both a fire hazard and failure of the circuit.

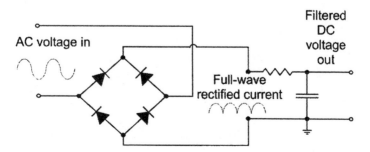

FIGURE 9.7 Rectifier with low-pass filter.

FIGURE 9.8 Heat sink.

fastfacts

Low-voltage DC wiring does not carry much current or power. However, if such signal or control wiring enters a high voltage box, local codes usually dictate that the temperature and voltage ratings on the wire must be the same as high voltage/high power wiring that enters that box.

AC CIRCUITS

The majority of large HVAC equipment is powered by alternating current (AC). The alternating part of this term is appropriate: the voltage and current in AC circuits flips back and forth 60 times per second. The primary service enters the building at 120/240 VAC if single phase, or 120/208 VAC, 277/480 VAC, and higher if three phase. The first number is the voltage between any single phase and ground and the second number is the voltage between any two phases. In commercial buildings, electricity enters a building though the primary service feeder (as shown in Figure 9.9) and then passes through a series of transformers, switchboards, and panels before arriving at the load. The HVAC equipment is usually powered directly from the switchboards. If a switchboard is devoted to just HVAC equipment, it is called a *motor control center*.

Most residential buildings are 120/240 VAC single-phase, three-wire systems. In this arrangement, a single-phase transformer is center tapped with this tap referenced to ground. The two hot wires from the transformer ends have 240 VAC, and each is 120 VAC with respect to ground. So plug loads can use the 120 VAC while larger loads such as the range, furnace fan, electric water heater, and central air heat pump can take advantage of the 240 VAC.

Most commercial buildings are three phase at either 120/208 VAC, 277/480 VAC, on up to 2400/4160 VAC. This latter is found only in the largest buildings, although some older commercial buildings still utilize high voltage HVAC equipment for the chillers and boilers.

fastfacts

Sometimes 120 V equipment is listed at 115 V, and 480 V equipment at 460 V. These differences are due to a standard 4% feeder voltage drop that is experienced almost everywhere.

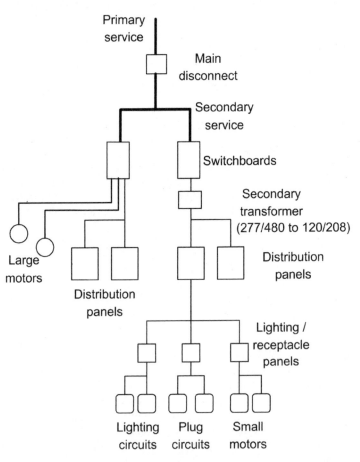

FIGURE 9.9 Typical commercial building electrical distribution system. *(Adapted from Stein and Reynolds)*

Ideally, all the varying current and voltage in an AC circuit would be in phase. That is, they would both reach their highest values at the same time. Since total power is voltage multiplied by current, this would give the highest power as well. However, as the electricity enters motors and other equipment, the voltage and current can get out of phase due to the presence of inductors and capacitors. Figure 9.10 shows what the voltage and current would look like if they were out of phase—in this case the current leads the voltage by 25 degrees out of the full 360 degrees required for one crest-to-crest cycle. The *power factor* is used to describe how much the current and voltage are out of phase. In this case, the power factor is the cosine of 25 degrees, or 90 percent. What this means is that only 90 percent of the possible power is going to run the device, while the remaining 10 percent does nothing. With large HVAC equipment, you want to keep your power factor as high as possible because sometimes the utility penalizes you if your building has a low power factor.

Wye and Delta Circuits

Large fan and pump motors use either a wye or delta wiring configuration, although delta configurations are much more common. The names of these circuits come from the connections of the motor windings as shown in Figure 9.11. The different configurations can be used to control the *inrush current;* that is, the amount of current

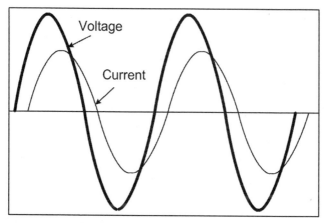

FIGURE 9.10 AC voltage and current 25° out of phase.

fastfacts

The electric grid in the United States uses 60 Hz power, but in other countries the grid operates at 50 Hz. If you use electric components manufactured overseas, make sure they are compatible with the electrical service you are using. Equipment specified for one frequency should not be used at another.

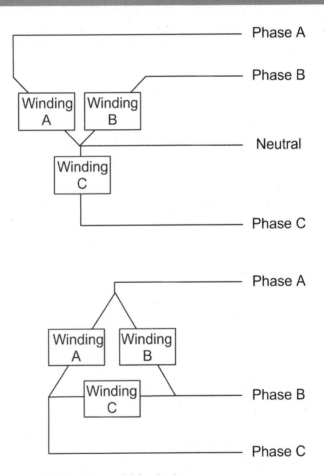

FIGURE 9.11 Wye and delta circuits.

that flows into the motor windings on startup. When starting a large motor, it is desirable to minimize this inrush current because too large a value can overheat the motor windings.

Variable Speed Drives

Motors spin at a speed related to the line cycle frequency. In the United States, the line cycle frequency is 60 Hz, so most motors operate at some multiple of this such as 1800 or 3600 RPM. However, often it is useful to have a motor operate at a different or variable speed. A clutch and gearing system can be used but is expensive and difficult to maintain. Since the late 1980s, *variable frequency drives* (VFDs or sometimes *variable speed drives*, VSDs) have become quite popular for applications that need changing motor speeds. Many variable air volume systems and large water distribution systems now use these drives, along with cooling towers and some kinds of chillers. VFDs change the actual speed of the motor by creating a new AC line voltage at any desired frequency. Figure 9.12 shows the principle behind the operation of a variable frequency drive. The

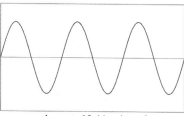

a. input 60 Hz signal

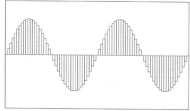

b. synthesized 40 Hz signal

FIGURE 9.12 Comparison of 60 Hz and 40 Hz signals.

standard line frequency (top) is decomposed and then used to generate a series of impulses that "blend together" to create a new periodic wave at a different frequency. Figure 9.12 shows how 60 Hz power is converted to a 40 Hz signal to slow a motor.

With VFDs it is important to remember that the motor may not be not operating at the design frequency, and that there may be certain frequencies that cause excessive vibrations or harmonics in the motor and attached fan or pump. Most VFDs have *critical speed step-over* circuits that are adjustable ranges that can be locked out to avoid any resonant speeds. Two step-over ranges of individually adjustable widths are sufficient for most applications. In addition, the VFDs can introduce noise back on to the *incoming* electrical lines that can affect the other motors in the building as well as sensitive electronic equipment. In many cases the VFDs are located near the motor control center (as in Figure 9.13) so that such problems can be easily traced. Identifying the culprit for such *harmonics* usually requires the use of an oscilloscope or other specialized equipment.

FIGURE 9.13 Motor control center (cabinet on left) and variable frequency drives (two boxes on right).

Transformers

Transformers are used to change the voltage in alternating current circuits. A *step-up* transformer increases the voltage, a *step-down* transformer decreases the voltage. Electromagnetic induction is used to transmit the energy from the input (primary) side of the transformer to the output (secondary) side of the transformer (Figure 9.14). Because the two windings of a transformer are not directly connected, the transformer provides isolation, that is, it protects the load from noise on the line and protects the line from noise produced by the load. An isolation transformer may have the same input and output voltage. Transformers may also have multiple windings to allow a single input to produce multiple output voltages.

Transformers come in a wide variety of sizes, from the very small (as used in control systems) to very large (to drop kilovolt line voltages down to the 480 or 120 VAC used in the building). Transformers are rated according to their maximum power-handling capacity. This value is given in volt-amperes (VA). A typical low voltage transformer used for an actuator might have a rating in the 100 VA range (Figure 9.15) while a large, whole-building transformer is in the 1000 kVA range or larger (Figure 9.16).

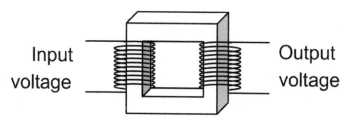

Input voltage   Output voltage

FIGURE 9.14   Transformer.

fastfacts

Transformers only work with AC voltages. Use a voltage divider on a DC circuit if you want to provide a stepped-down voltage to a sensor.

FIGURE 9.15 750 VA Transformer.

Most electronic valve and damper actuators use a dedicated 24 VAC or 48 VAC transformer for power. You should use a separate transformer for each actuator. There are two reasons for this, the first being that of reliability: if the transformer fails, you lose only one actuator rather than many that might be connected to that transformer. The other reason is a bit more obscure and has to do with floating versus grounded transformers. When working with transformers, remember that there is no physical connection between

fastfacts

If a modulating electronic actuator moves on its own or does not respond properly to the control signals, check to make sure that it has a floating transformer.

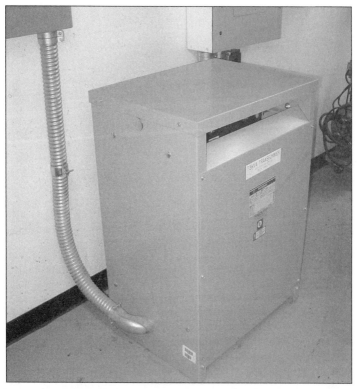

FIGURE 9.16   750 kVA Transformer.

the primary and secondary side. You can therefore reference the secondary to any arbitrary ground, or not attach it to ground at all. In the latter case, this is called a *floating* transformer. Figure 9.17 shows the difference between a grounded and floating circuit. Many actuators expect to see a floating circuit. If you power more than one actuator with a single transformer, each actuator will be referenced to the others and the circuit will not strictly be floating.

Note, however, that most large motor and fan secondary circuits are grounded, primarily for human and equipment safety reasons. In these cases, the ground should never be defeated by switches, circuit breakers, or fuses.

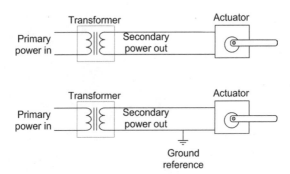

FIGURE 9.17 Floating (top) versus grounded (bottom) transformer circuits.

Circuit Breakers and Fuses

Practically all electronic equipment, whether DC or AC, is protected by circuit breakers, fuses, or some combination of the two. Circuit breakers are usually found in a breaker panel or a motor control center and can trip out when the current goes above a certain level or the breaker temperature gets too high. In both cases, the purpose of the breaker is to cut power to a device to prevent damage from too high a current flow. Breakers are generally resettable, meaning that you can close the circuit again once the breaker has tripped.

Some circuit breakers, particularly in residential applications, are *ground fault interrupts* (GFIs). These kinds of breakers measure not only the current heading to the appliance on the hot lead, but also the return current on the neutral. Any difference between the two implies that there is a path for the current other than the desired electrical circuit. If this occurs, the breaker will trip off. These kinds

fastfacts

Some thermal breakers are more susceptible to line noise. For example, variable frequency drives and unbalanced motors can cause repeated trip-outs of thermal breakers even though the actual current is not above the rated limit.

fastfacts

Fuses provide protection not only to equipment but to personnel as well. Never defeat fuse protection on a circuit. If a fuse continues to fail, identify the cause before returning the equipment to service.

of breakers are used especially in residential bathroom circuits where there are water, electricity, and occupants all together. However, you may also find GFI breakers on residential furnace fans and heat pumps that are meant to act as protection against any unintended short circuits.

Fuses are similar to breakers in that they halt the flow of current if it gets too high. The difference is that fuses must be replaced once they have broken the circuit. Fuses are located in both motor control centers as well as at the line input for most HVAC equipment. Figure 9.18 shows a fuse bank for an electric resistance heater in an air-handling unit. These fuses are rated at about 30 amperes. These fuses are classified as *slow-blow* fuses, meaning that there is some

FIGURE 9.18 Fuse bank.

delay between the occurrence of high current and the fuse failing. The reason for this is so that the fuse can tolerate brief periods of high inrush current to motors.

Electric Switches

Electric switches can be used for turning devices on and off, and for opening and closing valves and dampers. While the functions and sizes of the switches can differ, the principle is based on the energizing of a small electromagnet that is then used to close a set of contacts or to move a two-position actuator. The switches can be broken down into three different categories

A *relay* is used to switch AC and DC circuits up to about 120 V. The basic principle of a relay is shown in Figure 9.19, where the top portion of the figure shows the relay in a de-energized state and the bottom portion shows the energized version. The relay electromagnet (coil) is powered by a 24 or 120 VAC signal. While the figure shows a single contact between the terminals and common, in practice relays can have multiple contacts, or *poles*. This figure shows a *double-throw* relay in which connections are made in both the de-energized and energized states. A *single-throw* relay would not have two sets of contacts but instead would just have a NO or NC contact. The type of relay is often abbreviated; for example, a DPDT refers to a double-pole, double-throw relay that could connect two separate circuits, each with a NO and NC contact. Figure 9.20 shows a picture of a standard three-pole, double-throw relay. On these kinds of devices the switching mechanism can detach from the terminal base so that replacements can be made without having to remove the wires.

A balance relay (also known as a mouse trap relay) is a three-position switch used to run small actuator motors. One set of contacts drives the motor in one direction, the other drives the motor in the opposite direction, and the third is used to stop the motor. This kind of tri-state actuator is used on damper and valve motors. A *contactor* is basically a three-pole, single-throw relay that is used to switch high voltage multiphase equipment on and off. Contactors are used to activate pretty much all motors, fans, pumps, and compressors in HVAC applications. Figure 9.21 shows a picture of a three-phase contactor rated for 480 V service at 30 amps. One problem that can happen in contactors occurs just as the coil is energized. The contacts can hit each other and then bounce off before hitting again. Multiple bounces of the contacts is referred to as chattering. A spark can jump across the small gap between the contacts

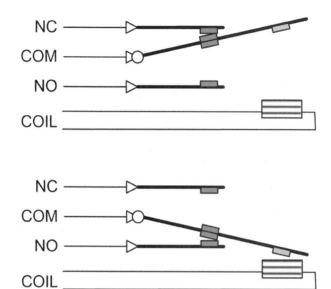

FIGURE 9.19 Schematic of electronic relay.

FIGURE 9.20 Electronic relay.

FIGURE 9.21 Three-phase contactor.

and too much chattering and can lead to eventual pitting of the contacts and poor electrical connections. This can, in turn, lead to an overheating of the contact and a failure of the contactor to either close that particular phase, or, in a worse case, to fuse the contacts so that the equipment remains powered even when then contactor is de-energized. For this reason, the contactor usually has internal springs that help make and keep the connection once the solenoid is energized.

Solenoids are two-position electromagnetic actuators that are used to open small valves on refrigerant lines, steam lines, and gas lines. They work in much the same fashion as contactors: a spring-return shaft is forced open by the energizing of a small coil. However, instead of closing electrical contacts, the electromagnet moves a rod of metal that can be used to close and open valves.

WIRES

Wire ratings are used to determine the size and current-carrying capacity of the wire. If you use wire that is too small for the amount of current, the wire will get too hot and will be a fire hazard. There are usually markings on the side of the wire jacket (Figure 9.22) that list information about the wire. In the United States, wire sizes are listed in American Wire Gauge with smaller numbers being larger wires as shown in Table 9.3. For large commercial HVAC systems you find wire sizes in 14, 12, 8, 6, 4, 3, 1, 0, 00, 000, and 0000 gauges. 14 gauge wire can carry between 25 and 35 amperes, depending on the insulation, while 0000 gauge wire can carry between 300 and 400 amperes.

The dampness ratings of cable are DRY, DAMP, and WET. In the first case, the cable is suitable for indoor applications above floor level where there is no chance the cable will get wet. DAMP rated wire should be used in locations such as interior locations near the floor, while WET rated cable should be used in all outdoor or buried wire runs.

Wire is also listed by a series of letters that describe the insulation, ampacity, and temperature ratings. Some of these ratings are listed in Tables 9.4 and 9.5. This allows you to identify some of the more common wiring types such as T, TW, THW, RHW, and RHH.

Residential HVAC systems are often wired with standard residential cable such as BX or ROMEX. When using these types of wiring, make sure to follow local residential codes. For example, the National Electric Code has a minimum of 12 gauge wire for residential use.

Certain wire color conventions are used in residential and commercial electrical systems. Follow these conventions when wiring single-phase HVAC systems to avoid confusion and to facilitate future troubleshooting. Typically the black wire is hot, the white wire is neutral, and the green wire is ground.

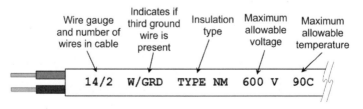

FIGURE 9.22 Example electrical cable and associated markings.

TABLE 9.3 Properties of Standard Copper Wire

Gauge, AWG	Ohms per 1000 feet	Diameter, inches	Weight, lbs/1000 feet
0000	0.05	0.460	641
000	0.07	0.410	508
00	0.09	0.365	403
0	0.11	0.325	319
1	0.14	0.289	253
2	0.17	0.258	201
3	0.22	0.229	159
4	0.27	0.204	126
5	0.34	0.182	100
6	0.43	0.162	79.5
7	0.55	0.144	63.0
8	0.69	0.128	50.0
9	0.87	0.114	39.6
10	1.1	0.102	31.4
11	1.4	0.091	24.9
12	1.7	0.081	19.8
13	2.2	0.072	15.7
14	2.8	0.064	12.4
15	3.5	0.057	9.86
16	4.4	0.051	7.82
17	5.6	0.045	6.20
18	7.0	0.040	4.92
19	8.8	0.036	3.90
20	11	0.032	3.09
21	14	0.0285	2.45
22	18	0.0253	1.94
23	22	0.0226	1.54
24	28	0.0201	1.22
25	36	0.0179	0.97

fastfacts

Never put BX or ROMEX wire inside a conduit. The wire can overheat and be a fire hazard.

TABLE 9.4 Prefixes Used in Wire Ratings

Prefix	Meaning
T	Dry only, maximum temperature of 140°F
CF	Fixture wire with cotton insulation and maximum temperature of 190°F
AF	Fixture wire with asbestos insulation and maximum temperature of 300°F
SF	Fixture wire with silicon insulation and maximum temperature of 390°F
R	Rubber covered
S	Extra hard service appliance cord
SJ	Lighter service appliance cord
SV	Light service appliance cord
SP	Lamp cord with rubber insulation
SPT	Lamp cord with plastic insulation
X	Moisture resistant synthetic polymer insulation
FEP	Fluorinated Ethylene Propylene insulation for dry service over 190°F

TABLE 9.5 Suffixes Used in Wire Ratings

Suffix	Meaning
B	Indicates outer braid of glass fiber or other material
H	Loaded current temperature rating up to 165°F
HH	Loaded current temperature rating up to 190°F
L	Lead jacket
N	Nylon or thermoplastic polyester jacket
O	Neoprene jacketed wire
W	Wet use type

In commercial HVAC applications, large current draws come from motors and compressors. Typical full load current from such devices can vary as shown in Table 9.6. These currents are given for loads at 460 VAC. If the device is running at 230 VAC then you need to double the values in the table. As the amount of current running through a wire increases, so does the temperature of the wire. For this reason the National Electric Code puts limits on how many wires can be run through a single conduit. These values are summarized in Tables 9.7 through 9.9 for standard conduit sizes.

TABLE 9.6 Full Load Current Draw From Motors of Different Sizes

Motor HP	Full load amps at 460 V
1	1.7
2	3.0
3	4.5
5	7.5
10	14
20	26
30	39
40	51
50	63
75	90
100	123
150	180
200	240
300	348

TABLE 9.7 Maximum Allowable Number of Conductors in Conduit

AWG	Conduit size in inches								
	½	¾	1	1¼	1½	2	2½	3	3½
6	1	2	4	7	10	16	NR	NR	NR
4	1	1	3	5	7	12	17	NR	NR
2	1	1	2	4	5	9	13	NR	NR
1	NR	1	1	3	4	6	9	NR	NR
0	NR	1	1	2	3	5	8	12	NR
00	NR	1	1	1	3	5	7	10	NR
000	NR	1	1	1	2	4	6	9	12
0000	NR	NR	1	1	1	3	5	7	10

TABLE 9.8 Maximum Allowable Number of Conductors in Conduit (THHN Wire Type)

AWG	Conduit size in inches							
	½	¾	1	1¼	1½	2	2½	3
6	1	4	6	11	15	26	NR	NR
4	1	2	4	7	9	16	NR	NR
2	1	1	3	5	7	11	16	25
1	NR	1	1	3	5	8	12	18

fastfacts

Always check local codes for any further restrictions on types of wire and allowable conduit wire density that can be used in your application.

TABLE 9.9 Maximum Allowable Number of Conductors in Conduit (XHHN Wire Type)

AWG	Conduit size in inches							
	¾	1	1¼	1½	2	2½	3	3½
0	1	1	3	4	7	10	15	NR
00	1	1	2	3	6	8	13	NR
000	1	1	1	3	5	7	11	14
0000	1	1	1	2	4	6	9	12

POWER MEASUREMENT AND ELECTRIC RATES

Electrical meters are used to measure both the instantaneous power consumption of equipment as well as the total energy use over time. Instantaneous power is measured in watts (W) or kilowatts (kW) and is analogous to water flow rate: a watt meter measures how many electrons flow into a device every second, while a turbine meter measures how many gallons of water flow into a device every minute. Total energy is the total number of electrons that have entered a device over a period of time. This is measured in watt-hours (Wh) or kilowatt-hours (kWh). To use the water comparison again, an electric meter shows how many kWh a device has used over the course of a month, while a water meter shows the total number of gallons that a device has used.

Knowing your electricity use is important for two reasons. The first, of course, is that the utility charges you for the amount of electricity you use (and also for the peak rate at which you use it). The second reason to measure electricity use is to see how well your equipment is running. For example, if you measure the electricity consumption of a chiller and compare that to the actual cooling produced by the chiller, you can get a good idea as to whether or not the chiller is operating correctly.

fastfacts

Instantaneous power in watts is a rate. Total energy in watt-hours is an amount.

There are many ways to measure the power consumption of chillers, fans, and boilers, ranging from the very simple up to the very complex and expensive. One of the easiest ways to measure power consumption is to use current transducers to determine the ampere draw of a device. Assuming that the voltage will not vary all that much, and also assuming that the legs of a multi-phase motor are pretty well balanced, you can use Tables 9.10 through 9.12 to find the power as a function of the current. This also works if you take spot measurements of the current using a clip-on ammeter. Note that to use the data in this table, it is necessary to know the power factor of the device you are working with. Unfortunately, it is not always easy to determine the power factor and you may have to purchase or rent a special meter for this.

The local electric utility charges for electricity use. The rate schedules used to calculate the bills can be complex. It is the goal of the building manager to choose rates that best fit the building energy use so that the operating costs are minimized. It is therefore important to understand the various components of utility rates since they directly affect the total monthly energy bill.

Unfortunately, there are thousands of electric and gas rates in the United States, each with its own structure and quirks. While many rates appear to be similar, the details of each can determine whether or not it is appropriate for a given building. For example, some utility

fastfacts

In the same way you can tell if there is a problem with your car by keeping track of the miles per gallon of gas, you can identify problems with HVAC equipment by tracking the electrical efficiency.

TABLE 9.10 Calculation of Power in Single Phase 120/240 Circuits

i = current and PF = power factor

Circuit Type	Power
1 phase, 2 wire, with neutral	i x 120 x PF
1 phase, 2 wire, no neutral	i x 240 x PF
1 phase, 3 wire, with neutral	I x 2 x 120 x PF

TABLE 9.11 Calculation of Power in Three Phase 120/208 Circuits

i = current and PF = power factor

Circuit Type	Power
1 phase, 2 wire, w/ neutral	i x 120 x PF
1 phase, 2 wire, no neutral	i x 208 x PF
1 phase, 3 wire, w/ neutral	I x 2 x 120 x PF
3 phase, 3 wire, Delta	i x 1.732 x 208 x PF
3 phase, 4 wire Wye	i x 1.732 x 208 x PF
3 phase, 4 wire Delta	i x 1.732 x 240 x PF

TABLE 9.12 Calculation of Power in Three Phase 277/480 Circuits

i = current and PF = power factor

Circuit Type	Power
1 phase, 2 wire, w/ neutral	i x 277 x PF
1 phase, 2 wire, no neutral	i x 480 x PF
1 phase, 3 wire, w/ neutral	i x 2 x 277 x PF
3 phase, 3 wire Delta	i x 1.732 x 480 x PF
3 phase, 4 wire Wye	i x 1.732 x 480 x PF

rates charge more for energy in the middle of the day than at night. While the *average* rate is low, the coincidence of the high energy prices with high daytime use can be very expensive. It may be better to find a rate that has a higher average price but does not penalize for the on-peak periods.

The overall use of electricity (in kWh) or gas (in Btu or therms) over a given interval of time is called the total *consumption*. The consumption charge can change depending on the total consumption up to the current time within a given billing period, or it can change depending on the time of day. The highest instantaneous use of electricity (in kW) at a given interval of time is called the *demand*. Like the consumption, the cost of demand can depend on the time of day.

Utility rates are made up of consumption, demand, surcharge, adjustment, meter charges, and tax components. In addition, there are riders and provisions that may affect the rate or even if the building or campus can use a particular rate. In addition, the various consumption and demand costs can change from season to season or based on whether or not it is a weekday, weekend or holiday.

Consumption

Fixed fee consumption charges are typical for most residential bills. These kinds of charges are very simple: the charge is the same throughout the billing period and throughout the day, regardless of the time of day or how much energy has already been used. For electricity rates this corresponds to a fixed fee per kWh consumed.

Block rates vary the cost of electricity depending on how much electricity is used. The first block has a given size and cost. When that block is "filled" you move on to the next block, and so on. Figure 9.23 shows an example of a declining block rate, where the more electricity you use, the lower the unit cost becomes. Many commercial electricity rates are block rates.

Time-of-use or *time-of-day* rates vary the cost of electricity at each hour of the day as illustrated in Figure 9.24. The highest rates are typically from noon through 3 p.m., corresponding with high electricity use for cooling. Many commercial electricity rates incorporate some degree of time-of-use billing. The number of time-of-use periods can vary from two (on-peak and off-peak) up to five.

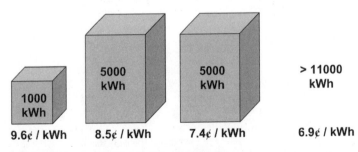

FIGURE 9.23 Example of declining block rate.

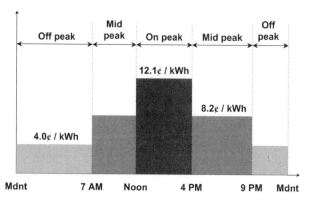

FIGURE 9.24 Example of time-of-use rate.

A large number of commercial and industrial electricity rates are combined block and time-of-use rates. For example, the on-peak period may have a block component so that there is block pricing structure that applies between certain hours of the day. Table 9.13 shows an example of a combined rate structure. Other adjustments on basic consumption charges include day type adjustments (where weekdays and holidays are metered at a different rate than weekdays) and seasonal adjustments where there are summer and winter differences, or four seasonal changes, or even cases where the rate structure changes each month.

Real-time pricing is a relatively new method available to large customers for electricity. In real-time pricing, the electricity rate changes every hour of the day depending on the utility's cost for producing the electricity. While the cost is generally very inexpensive most of the time, there are about 40 to 80 hours per year when

TABLE 9.13 Example Combined Time-of-Use and Block Electricity

Period	Cost	Comments
On-peak	9.362¢ for first 2500 kWh 12.48¢ for next 5000 kWh 15.66¢ for anything above 7500 kWh	On peak is 9:30 AM - 4:30 PM Monday through Friday except holidays.
Off-peak	5.763¢ for first 1700 kWh 6.311¢ for next 1300 kWh 8.210¢ for anything above 3000 kWh	All other periods are off-peak.

the cost can increase by a factor of ten or twenty. In these kinds of price structures, there is typically only a consumption charge with no demand charge.

Demand

The demand charges have block and time-of-use rates very similar to those used in consumption charges. The demand over a given billing period is usually set by the highest rate of energy consumption during any 15 minute period, although sometimes 30 minute or 60 minute intervals are used. The demand charges for a billing period are set by the highest demand *at any point* during that billing period. For example, Figure 9.25 shows the hourly demand values for a large office building where the monthly demand is set on the first Monday. Even though the rest of the month has a lower demand value, the monthly energy bill uses that peak demand for the calculation of the demand charge. Some energy rates have both on-peak and off-peak demand charges.

In some cases, the demand charges are *ratcheted* from month to month. This means that the peak demand charge can carry over from one month to the next, and sometimes even over the next year. There are many variations on this, such as having the demand charge for a given month be the greater of either the measured demand for that month or some percentage of the peak demand for any of the previous 11 months.

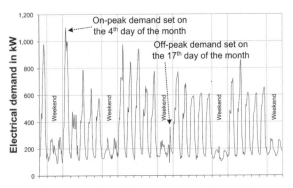

FIGURE 9.25   Example of monthly demand values and demand charge values.

For the HVAC engineer, an understanding of the demand rate used for a given building can play a big role in deciding what kind of energy conservation measures should be used. The cost of electricity demand can be almost equal to the cost of electricity consumption, and while it is difficult to reduce total consumption, it is somewhat easier to reduce the peak demand. For this reason, many energy saving techniques refer to *peak-shaving*, which tries to limit the peak demand for on-peak periods. Many larger chillers (see Chapter 8) have built-in controls that allow the user to set an upper limit on the kW use by the chiller compressor. Other pieces of HVAC equipment, such as thermal energy storage systems (see Chapter 10) are meant exclusively to perform peak shaving.

Cost Adjustments and Other Charges

On top of the consumption and demand charges, the utilities often apply cost adjustments that reflect their costs of producing or transporting the energy from the power plant. For electricity bills this is usually listed as the *energy cost adjustment* and is given in as additional cost per kWh or as a percentage of the total consumption and demand bill components. The cost adjustments can be positive or negative, depending on whether or not the cost to the utility for providing the service is more or less than expected.

Surcharges are additional fees applied to your bill that reflect charges applied for a variety of reasons, many of which are not directly related to your energy use. For example, many states impose an energy assistance surcharge that is used for a fund that helps subsidize low-income residential energy bills. Other surcharges come from exploration fees (often seen on your bill as "energy procurement") and energy transport fees ("interstate transition"). The surcharges can be a fixed fee per month, a function of total consumption, or a percent of the total bill.

Meter charges are seen in just about every energy bill, both gas and electric. The meter charges reflect the cost of installing and reading the meter every month. These charges can vary from several dollars per month for residential meters up to several hundred dollars per month for commercial and industrial customers. Generally, the more complex your rate, the more expensive the meter and the higher your monthly meter charge. For example, if you have a fixed consumption fee then the electrical meter can be relatively simple and cheap. But if you are on a time-of-use rate then the meter must be able to account for consumption at different times of the day and will be a more expensive meter.

Local, county, state, and federal *taxes* are all applied to various bill components. The total taxes can add up to anywhere from nothing to about ten percent of total bill. Finally, *riders* and *special provisions* are rules applied to each rate that dictate the use of the rate by a certain customer, or can provide incentives or added costs depending on the type of customer or the maximum load.

10

MATERIALS

The materials used by electricians are numerous. There are various options to choose from within each category. For example, the selection of proper staples or straps can require some thought. Deciding on what type of box to use and whether it should be metal or plastic is a consideration. Choosing the right materials can make a job easier to do and less expensive to complete. Article 110 of the 2002 N.E.C. requires that listed or labeled equipment be installed according to any instructions that are included with that listing or labeling. This is telling you that products should only be used for the intended purpose. It goes hand-in-hand with the old adage, "Use the right tool for the right job." You would not use a wrench as a hammer; similarly, you should not use a cable staple to secure conduit. With this in mind, let's explore materials that are commonly used in the electrical trade.

RECEPTACLES

Receptacles are used extensively in the electrical trade. Duplex receptacles in both 15 amp and 20 amp versions are commonly used in new construction, remodeling, and service work (Figure 10.1).

FIGURE 10.1 Duplex receptacle.

Specification Grade Receptacles

Specification grade receptacles are most often used in commercial applications. These receptacles are also used in industrial jobs. Some of the common locations to find specification grade receptacles include schools, offices, and grocery stores.

Clock Receptacles

Clock receptacles are used in some special cases. These receptacles are recessed in a box and allow a clock to be hung over the receptacle with room for the clock to be plugged into the outlet.

fastfacts

When installing a receptacle, you should match the amp rating of the receptacle with the size of the circuit. However, in multi outlet circuits, 15 amp receptacle may be installed on a 20 amp circuit.

Did You Know: that switches and receptacles that are designed for use with aluminum wire are marked CO/ALR? If you discover aluminum wire attached to a switch or receptacle that is not marked with CO/ALR, replace the device with an approved device.

230 VOLT RECEPTACLES

230 volt receptacles are available in 15 to 50 amp ratings. The receptacles are configured so that only a particular style of cord end will fit the receptacle (Figures 10.2 through 10.5).

30 Amp Dryer Receptacle

A 30 amp dryer receptacle can come in many styles. The receptacle might be a flush-mount unit or a surface-mount outlet (Figures 10.6, 10.7). This type of receptacle is designed to accept only plug ends used with clothes dryers. Three wire dryer receptacles are now allowed only for replacement of existing receptacles. All new 30 amp dryer receptacle installations must be four wire.

50 Amp Range Receptacle

A 50 amp range receptacle can come in many styles. The receptacle might be a flush-mount unit or a surface-mount outlet (Figures 10.6, 10.7). This type of receptacle is designed to accept only plug ends used with electric ranges used in kitchens. Three wire range receptacles are allowed only for replacement of existing receptacles. All new 50 amp receptacle installations must be four wire.

 Trade Tip: A #6 wire is rated for 60 amps and 240 volts. This type of wire is used for central air conditioners, electric furnaces, and similar equipment.

Did You Know: that many plastic boxes are labeled with the maximum number of wires in various gauges that are allowed in the box? Many boxes do offer this labeling to enable an electrician to avoid overloading the box.

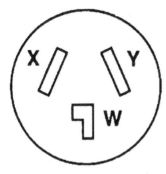

FIGURE 10.2 30 A 125/250 V 3 pole, 3 wire receptacle.

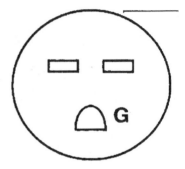

FIGURE 10.3 30 A 250 V grounding 2 pole, 3 wire receptacle.

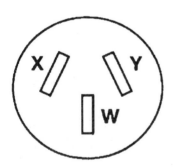

FIGURE 10.4 50 A 125/250 V 3 pole, 3 wire receptacle.

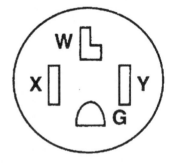

FIGURE 10.5 30 A 125/250 V grounding 3 pole, 4 wire receptacle.

Trade Tip: Inspect the insulation on all wiring once you are ready to make a connection. Be sure that the insulation is not damaged.

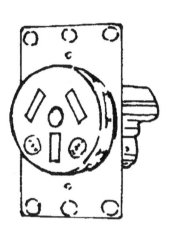

FIGURE 10.6 Flush mount receptacle.

FIGURE 10.7 Surface mount receptacle.

Twist-Lock Receptacles

Twist-lock receptacles are available in many styles and with various voltage ratings. Amp ratings for these devices range from 15 amps to 60 amps. Motors are the most common application for twist-lock receptacles. This type of receptacle can be found on generators, welders, and so forth. A 20 amp twist-lock male cord end coupled with a 20 amp twist-lock female end helps to make a very user-friendly extension cord for electricians. This type of installation can be used with normal electrical wire to create an extension cord of most any length. Plug the extension cord into the GFI receptacle at a temporary power pole site and you are ready to rough-in a job.

Isolated Ground Receptacles

Isolated ground receptacles are orange in color or they can be another color with a small orange triangle on their face. They are designed so that the ground prong is isolated from building steel. An insulated green wire goes directly back to the ground bus in the panel board. The frame or yoke of the receptacle is bonded to the normal equipment ground of the system. They are used where electrical interference or noise on the ground wire is a problem. This usually occurs with computers and other sensitive electronic equipment. An outlet of this type can be grounded separately from the normal grounding circuit. The isolated ground eliminates electromagnetic interference through a separate grounding path. This grounding path isolates the transient current that can cause electronic equipment, such as cash registers, computers, and hospital equipment, to perform poorly.

GFI Receptacles

GFI receptacles are available in residential, commercial, and hospital grades. These receptacles can be feed-through types or non-feed-through types. Feed-through types protect receptacles further down the circuit. A set of load terminals exist on the backs of these types of GFI receptacles. One terminal is white and the other is brass. Additional receptacles farther down a circuit are wired to the load terminals. This protects the outlets downstream of the GFI receptacle with the load wires.

Non-feed-through GFI receptacles protect only the primary outlet. Receptacles downstream of a non-feed-through GRI receptacle do not benefit from added protection. The basic features of both feed-through and non-feed-through GFI receptacles are:

- *They trip at 5 MA threshold*
- *They trip in 0.025 seconds for a 240 MA fault*
- *They have test buttons*
- *They have reset buttons*

Some GFI receptacles have indicator lights as a part of the receptacle. However, not all GFI outlets have the indicator lights (Figure 10.8). The indicator light, when present, indicates that the receptacle is online. If a receptacle were tripped, the indicator light would be off. GFI receptacles that are not equipped with indicator lights have to be tested manually to see if they are online.

GFI cord sets and extension cords are also available. The cord sets come in lengths of 2 feet, 6 feet, and 25 feet. GFI extension cords come in a 37 foot length. These cord sets and extension cords comply with the Occupational Safety and Health Act (OSHA). These cord sets and extension cords are valuable on job sites and other locations where a permanent installation of a GFI circuit is not feasible. The cords mentioned here provide protection from both ground faults and open neutral conditions. Cord sets are enclosed in water-resistant protection. A power indicator light goes off when the device is tripped or the line cord is unplugged. Test and reset features on the devices permit instant confirmation that protection is available.

Commercial-Grade Designer Receptacles

Commercial-grade designer receptacles are offered in many colors and styles. These receptacles can be back wired or side wired. Terminal wires on commercial receptacles will accept up to 10 gauge

FIGURE 10.8 GFI receptacle.

wire. The back-wire holes in the receptacles will accept either 14 gauge wire or 12 gauge wire that is copper or copper-clad.

Commercial receptacles are fitted with labor saving self-grounding straps. Double wide brass contacts that are rated as heavy duty for a longer service life are also standard on these receptacles. The receptacles come in both 15 amp and 20 amp versions for 120 volts. There is also a model available that provides a 15 amp rating at 250 volts.

Residential-Grade Designer Receptacles

Residential-grade designer receptacles are available with a 15 amp rating at 120 volts. The receptacles come in many colors, such as mahogany, gray, almond, black, white, and ivory. These outlets can be back wired or side wired, and are self-grounding when installed in a properly grounded metal box. Heavy-duty, double wide brass contacts exist for longer service life.

Surge-Suppressor Receptacles

Surge-suppressor receptacles are state-of-the-art in style and protection. These receptacles protect sensitive electronic equipment, such as computers, from transient voltage surges. There are three levels of protection when a surge-suppressor receptacle is used. A monitor or indicator light shows when the three levels of protection are active. These receptacles are available with ratings of either 15 amps or 20 amps.

Surge-suppressor receptacles have an impact-resistant face and fuse-protected circuitry. Additionally, they have RFI and EMI noise filtration. Residential and commercials jobs can benefit from the installa-

fastfacts

Outlets are equipped with screws to place wires under. Many outlets have holes in the back of the receptacle so that stripped wire can be inserted into the hole for a connection. Few professionals use the stab method of connection. When wires are secured under screws, the connection is much more likely to maintain its integrity.

fastfacts

How many connectors are allowed to be secured under the screws of an outlet? Only one conductor is allowed. So, what do you do when bringing older wiring up to safer conditions? Use a jumper. Put one end of the jumper wire under the outlet screw. Then connect the other wires to the opposite end of the jumper and secure it with a wire nut. This is a much safer procedure.

tion of surge-suppressor receptacles in selected locations. Whenever there is a risk of a sudden surge of transient power that could damage sensitive equipment, a surge-suppressor receptacle should be installed.

SWITCHES

Switches, like receptacles, come in a wide variety of styles, types, and colors. Different voltage ratings are available for switches. Single-pole switches with ratings of 15 amps and 20 amps are the most common type of switch used in new homes, remodeling, offices, garages, and so forth. Three-way switches in the same amp ratings are also very popular in these job applications.

Switches are used to control lighting loads, but they can also be used to control small motors, such as those found on sump pumps, garbage disposals, exhaust fans, and so forth. Switches come in push-on-push-off configurations. They are also available in rotary styles.

I was wiring a school a few years ago and had an interesting experience with switches. There were four of us reviewing the plans and wiring diagrams for the job. The discussion was on how to wire the three-way and four-way switches in the school. In passing, I jokingly made a comment about them not forgetting the five-way switches. I assumed that these green electricians knew there was no such thing as a five-way switch. I was, however, rudely awakened when the job foreman told me not to confuse his crew so they could get some work done. Much to my surprise, the other three electricians had wasted the entire afternoon trying to locate the five-way switches on the plans.

Single pole and three-way switches are available in many styles. For example, you can get a switch that has two single pole switches. There is a switch that offers a single pole with a grounded receptacle. A single pole switch is available with a pilot light. You can even get a single pole switch and a three-way combination switch in a single unit. If you want a switch that gives you two three-way switches in one unit, you can get it.

There is also a three-way switch with a grounded receptacle. Combination devices, like those just discussed, all come in handy from time to time. When might you use these combination devices? Here are a couple of examples:

- *A switch with a pilot light can be used to indicate when the switch is in the on position. This can be very handy in certain applications. Have you ever been to a medical facility and seen a lighted switch on the wall outside of the bathroom door? This allows users to know that someone has the bathroom light on and that the bathroom is probably occupied.*

- *There may be a time when you will not have room to squeeze two switches into available space. If this happens, using a combination device will solve the space problem.*

Porcelain Light Fixtures

Porcelain light fixtures are generally provided with a pull-cord switch. This is a common setup, but there are options. Some porcelain light fixtures come with a convenience outlet on the flange of the light fixture. These fixtures are most often used in basements, attics, crawlspaces, and storage areas. While these fixtures are usually operated with a pull-chain switch, the fixture can be wired to work with other types of switches.

Lamp Switches

Lamp switches come in a myriad of fashions. Some of these types of switches include:

- *Pull-chain switches*
- *Turn-knob switches*
- *Key switches*
- *Keyless switches*
- *Push-type switches*

When faced with repairing lamp switches, you may be able to rebuild the defective switch. It is often easier to simply replace the defective device.

Dimmer Switches

Dimmer switches are available as single-pole switches and as three-way switches. The purpose of a dimmer is to dim incandescent lighting. These switches are used in homes, showrooms, display windows, and many other applications (Figure 10.9). Fluorescent dimmers are used to dim lighting in showrooms, displays, windows of commercial establishments, and similar areas.

ROCKER SWITCHES

Rocker switches combine style and rugged performance. A rocker switch can be wired from the back or the side of the switch. The

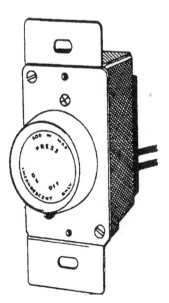

FIGURE 10.9 Push-pull dimmer switch.

concave rocker design provides a low profile and very quiet switching. Thermoplastic construction of rocker switches resists hard impacts and long-term abuse.

A one-piece, heavy metal strap with locking tabs is used in rocker switches to guarantee additional strength in the switch. Both solid and standard wire can be connected to a rocker switch. There are dual grounding connections on rocker switches. This gives you an option of a green grounding screw or a self-grounding clip that is riveted to the back strap of the switch. Some of the available types of rocker switches include:

- *Single pole*
- *Double pole*
- *Three way*
- *Four way*
- *Double throw with center off*
- *Double pole, double throw, center off*

Rocker switches come in a multitude of colors, such as ivory, white, and almond. Switch covers for these switches are available in matching colors. The cover plates fit tight and shield the switch from dust. A tight fit ensures that the rocker switch will not bind during use.

Approved uses of rocker switches include tungsten, florescent, and resistive loads. If used to switch a motor, the motor capacity should be no more than 80 percent of the switch rating.

SPECIALTY SWITCHES

There are some specialty switches that you may want to use occasionally. One such switch is a programmable switch. This type of switch is approved only for incandescent use. The switch provides a manual on and off function, but it also offers a programmable feature. The programmable feature allows the switch to work automatically, based on the programming done on the switch.

For example, the switch can be used to turn a light on and off when the switch is unattended. A battery-back-up feature is common in programmable switches. The battery allows the switch to maintain its program memory for about 72 hours. Wattage ratings for this type of switch range from 40 watts to 500 watts. Programmable switches are available as single-pole units and as three way switches.

Time-Delay Switch

Time-delay switches are limited to incandescent use only. These switches can be installed in a standard single gang wall box in place of a standard wall switch. The time-delay switch functions with an electronic keypad. Pigtail leads on the switch allow for fast and easy wiring. Common places to install time-delay switches include:

- *Garages*
- *Children's rooms*
- *Basements*
- *Stairways*
- *Hallways*
- *Walkways*

Time-delay switches are rated for 300 watts at 120 volts. When lights are on, a person can push the delayed switch button and have about five minutes before the light is turned off.

PLUGS

Straight blade plugs and connectors are available in 15 amp and 20 amp ratings. The residential grade is the only type available. There are also 10 amp and 15 amp versions available as parallel plugs and connectors that are used to repair cords that are polarized, but not grounded. 15 amp, 2 pole, 3-wire grounding plugs and connectors are used on appliance cords, electrical tools, extension cords, and for the replacement of damaged cord ends (Figure 10.10).

Industrial Grade Plugs

Industrial grade plugs come rated for 15 amps and 20 amps. These plugs are considered to be the top of the line when it comes to plug selection. A cable clamp is built into these plugs and connectors.

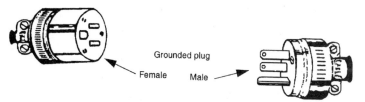

Grounded plug

Female   Male

FIGURE 10.10   Grounded plugs.

This minimizes cord damage when installing the cord end. The wire clamps are individual chambers that provide positive crimp locking without cutting the wire strands. Three fast-threading screws are used to secure the device once the wiring work is complete.

Industrial grade plugs are put through a battery of tests. They are certified for hospital use and can assure top performance in the harshest industrial environment. For example, the plugs are put through a 500 pound crush test. The plug is placed between two steel plates and squeezed to a point that is equal to 500 pounds. The test is done with a force rating of 2,500 pounds to simulate the worst abuse a plug might encounter when in use.

An impact test is also done on industrial grade plugs. This test consists of a ten pound weight being dropped repeatedly on a plug from a height of 18 inches.

Plugs on jobsites can be subject to a lot of abuse. The plugs can be stepped on, run over with equipment or staging wheels, and so forth. This means that the plugs have to be tough.

In a mechanical drop test, a cord is attached to a length of rubber cord with a plug or connector end made up on it. The plug or connector is pulled up and dropped from a height of about 45 inches and released so that the plug or connector will impact a hard, wooden surface. This test may be conducted hundreds of times.

A cord test is also done on industrial grade plugs and connectors. This test is done by making up a cord end on a length of rubber cord. The cord is placed in a holder and 30 pound straight pulls are performed on the cord. Cord clamps must remain securely fastened to the plug. Then, rotating pulls are done on the cord end with 10 pounds of weight with a circle motion for two hours. Once again, the cord clamp is required to remain solidly attached.

WALL PLATES

Wall plates are available in a wide variety of styles, colors, and types of material construction. Sizes ranging from standard size to jumbo size are available for wall plates. The construction materials include the following:

- *High-abuse construction*
- *Nylon*
- *Plastic*
- *Brass*
- *Aluminum*
- *Stainless steel*
- *Chrome*

WIRE CONNECTORS

Wire connectors are produced in configurations to match virtually every wiring application. Be certain that the connectors that you are using are rated for the application at hand. Connecting aluminum wiring to copper wiring can present a potential hazard if the connection is done improperly. One of the reasons for this is that when the insulation is stripped from the wire, a film immediately forms on the wire that can make for a poor and dangerous connection. Make sure you use only Al/Cu rated wire connectors when making these connections (Figures 10.11a, b).

Wire connectors are packaged with clearly labeled tables that list the various wire size combinations that can be used with those connectors. It's important to read and understand the limitations of wire connectors. Don't ever be tempted to use a wire connector smaller than the rated application. Doing so would be a code violation and make you subject to civil liabilities. Most importantly, you would be creating a life threatening hazard. Figures 10.12 through 10.15 illustrate a variety of wire connectors and corresponding tables.

WIRE SIZES AND APPLICATIONS

Using the correct wire size rated for a specific job is important to safe, trouble free wiring. Figures 10.16 through 10.33 illustrate examples of commonly used wire, cable, and cord sizes and applications.

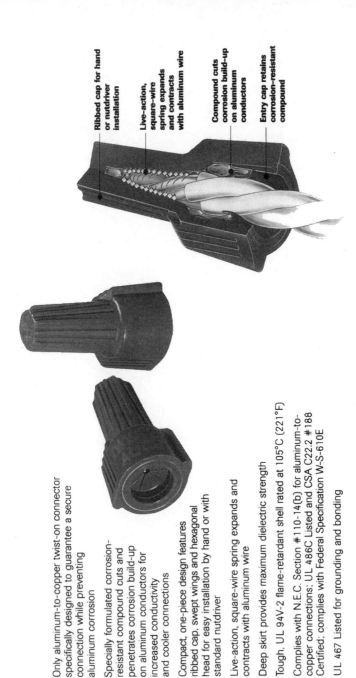

Labels on figure:
- Ribbed cap for hand or nutdriver installation
- Live-action, square-wire spring expands and contracts with aluminum wire
- Compound cuts corrosion build-up on aluminum conductors
- Entry cap retains corrosion-resistant compound

- Only aluminum-to-copper twist-on connector specifically designed to guarantee a secure connection while preventing aluminum corrosion

- Specially formulated corrosion-resistant compound cuts and penetrates corrosion build-up on aluminum conductors for increased conductivity and cooler connections

- Compact, one-piece design features ribbed cap, swept wings and hexagonal head for easy installation by hand or with standard nutdriver

- Live-action, square-wire spring expands and contracts with aluminum wire

- Deep skirt provides maximum dielectric strength

- Tough, UL 94V-2 flame-retardant shell rated at 105°C (221°F)

- Complies with N.E.C. Section #110-14(b) for aluminum-to-copper connections; UL 486C Listed and CSA C22.2 #188 Certified; complies with Federal Specification W-S-610E

- UL 467 Listed for grounding and bonding

FIGURE 10.11a Twister® Al/Cu wire connector. (*Courtesy Ideal Industries*)

Model	Color	Wire Combinations	Quantity	Cat. No.
65	Purple	1 No. 10 Al solid w/1 or 2 No. 10 Cu solid 1 No. 10 Al w/1 or 2 No. 12 Cu 2 No. 10 Al solid w/1 No. 12 Cu 1 No. 10 Al w/1 or 2 No. 14 Cu 2 No. 10 Al solid w/1 No. 14 Cu 2 No. 12 Al solid w/1 No. 10 Cu 1 No. 12 Al stranded w/1 or 2 No. 10 Cu solid	Card of 2	**30-065**
		1 No. 12 Al w/1 or 2 No. 12 Cu 2 No. 12 Al solid w/1 No. 12 Cu 1 No. 12 Al w/1 or 2 No. 14 Cu 2 No. 12 Al solid w/1 No. 14 Cu 1 No. 10 Al w/1 or 2 No. 18 Cu 2 No. 10 Al solid w/1 No. 18 Cu	Card of 10	**30-765**
		1 No. 12 Al w/1 or 2 No. 16 Cu 2 No. 12 Al solid w/1 No. 16 Cu	Card of 25	**30-165**
		1 No. 10 Al w/1 or 2 No. 16 Cu 2 No. 10 Al solid w/1 No. 16 Cu	Box of 100	**30-265**
		1 No. 12 Al w/1 or 2 No. 18 Cu 2 No. 12 Al solid w/1 No. 18 Cu	Carton of 1,000	**30-365**

600V maximum building wire; 1,000V signs or lighting fixtures

Wire Connector also listed as grounding equipment

FIGURE 10.11b Twister® Al/Cu wire connector, wire size combinations. *(Courtesy Ideal Industries)*

- Smallest wet location/direct burial connector with widest wire range – covers wire sizes from 20 AWG through 12 AWG

- Non-hardening sealant completely seals out moisture to protect conductors from fungus and corrosion – remains stable from -40°C (-40°F) to 105°C (221°F)

- Compact, one-piece design features ribbed cap, swept wings and hexagonal head for easy installation by hand or with standard nutdriver

- Live-action, square-wire spring locks onto wire for safe, secure connections

- Deep skirt provides maximum dielectric strength

- Tough, UL 94V-2 flame-retardant shell rated at 105°C (221°F)

- Certified for underground and wet locations – UL 486D Listed and CSA C22.2 #188.2 Certified; complies with Federal Specifications W-S-610E

Callouts on image: Ribbed cap for hand or nutdriver installation; Square-wire spring; Non-hardening sealant; Swept wings for added torque and comfort

Model	Color	Wire Combination Range	Wire Combination Range (mm)	Quantity	Cat. No.
60	Blue	600V* No. 22 to 8 AWG Min. 1 No. 18 & 1 No. 20 Max. 1 No. 10 & 2 No. 12	600V* .64mm DIA to 3.26mm DIA Min. 1.34mm² Max. 11.88mm²	Card of 2	30-060
				Card of 10	30-760
				Card of 25	30-160
				Box of 100	30-260
				Carton of 1,000	30-360

*1,000V maximum in fixtures and signs

FIGURE 10.12 Twister® DB Plus™ wire connector. *(Courtesy Ideal Industries)*

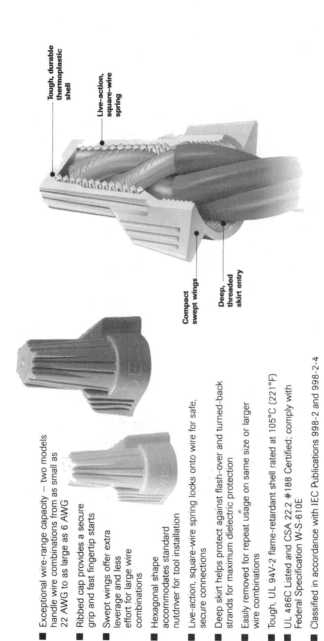

- Tough, durable thermoplastic shell
- Live-action, square-wire spring
- Compact swept wings
- Deep, threaded skirt entry

■ Exceptional wire-range capacity – two models handle wire combinations from as small as 22 AWG to as large as 6 AWG

■ Ribbed cap provides a secure grip and fast fingertip starts

■ Swept wings offer extra leverage and less effort for large wire combinations

■ Hexagonal shape accommodates standard nutdriver for tool installation

■ Live-action, square-wire spring locks onto wire for safe, secure connections

■ Deep skirt helps protect against flash-over and turned-back strands for maximum dielectric protection

■ Easily removed for repeat usage on same size or larger wire combinations

■ Tough, UL 94V-2 flame-retardant shell rated at 105°C (221°F)

■ UL 486C Listed and CSA 22.2 #188 Certified; comply with Federal Specification W-S-610E

■ Classified in accordance with IEC Publications 998-2 and 998-2-4

FIGURE 10.13a Twister® wire connectors. (*Courtesy Ideal Industries*)

203

Model	Color	Wire Combination Range	Wire Combination Range (mm)	Quantity	Cat. No.
341®	Tan	600V* No. 22 thru 8 AWG Min. 3 No. 22 Max. 3 No. 10	600V* .64mm DIA to 3.26mm DIA Min. .97mm² Max. 15.78mm²	Box of 50	**30-141**
				Box of 100	**30-341**
				Jar of 750	**30-341J**
				Bag of 500	**30-641**
				Carton of 1,000	**30-541**
342®	Gray	600V* No. 18 thru 8 AWG Min. 3 No. 14 Max 2 No. 10 w/4 No. 12 Solid	600V* 1.02mm DIA to 3.26mm DIA Min. 6.24mm² Max. 23.76mm²	Box of 25	**30-142**
				Box of 50	**30-342**
				Bag of 250	**30-642**

*1,000V maximum in fixtures and signs

UL LISTED WIRE-CONN   CSA   CU/CU Only   CE   EN 60-998-2-4

FIGURE 10.13b Twister® wire connector, wire size combinations. *(Courtesy Ideal Industries)*

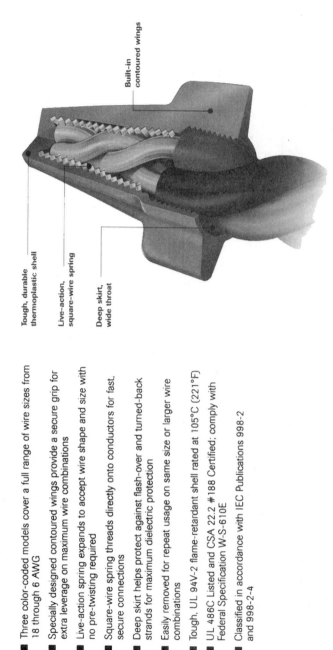

Built-in contoured wings

Tough, durable thermoplastic shell

Live-action, square-wire spring

Deep skirt, wide throat

- Three color-coded models cover a full range of wire sizes from 18 through 6 AWG
- Specially designed contoured wings provide a secure grip for extra leverage on maximum wire combinations
- Live-action spring expands to accept wire shape and size with no pre-twisting required
- Square-wire spring threads directly onto conductors for fast, secure connections
- Deep skirt helps protect against flash-over and turned-back strands for maximum dielectric protection
- Easily removed for repeat usage on same size or larger wire combinations
- Tough. UL 94V-2 flame-retardant shell rated at 105°C (221°F)
- UL 486C Listed and CSA 22.2 #188 Certified; comply with Federal Specification W-S-610E
- Classified in accordance with IEC Publications 998-2 and 998-2-4

FIGURE 10.14a Wing-Nut® wire connectors. *(Courtesy Ideal Industries)*

Model	Color	Wire Combination Range	Wire Combination Range (mm)	Quantity	Cat. No.
451®	Yellow	600V* No. 18 thru 10 AWG Min. 2 No. 18 Max. 3 No. 12	600V* 1.02mm DIA to 2.59mm DIA Min. 1.64mm² Max. 9.93mm²	Box of 100	30-451
				Jar of 225	30-451J
				Carton of 1,000	30-551
				Keg of 10,000 (20 bags, 500 ea.)	30-651
452®	Red	600V* No. 18 thru 8 AWG Min. 2 No. 14 Max. 5 No. 12	600V* 1.02mm DIA to 3.26mm DIA Min. 4.16 mm² Max. 16.55 mm²	Box of 100	30-452
				Jar of 300	30-452J
				Carton of 1,000	30-552
				Keg of 5,000 (10 bags, 500 ea.)	30-652
454®	Blue	600V* No. 14 thru 6 AWG Min. 3 No. 12 Max. 2 No. 6 & 1 No. 12	600V* 1.63mm DIA to 4.12mm DIA Min. 9.93mm² Max. 29.9mm²	Box of 50	30-454
				Keg of 2,500 (25 bags, 100 ea.)	30-654

*1,000V maximum in fixtures and signs

WIRE CONN
UL LISTED CuCu Only ® CE EN 60-998-2-4

FIGURE 10.14b Wing-Nut® wire connector, wire size combinations. *(Courtesy Ideal Industries)*

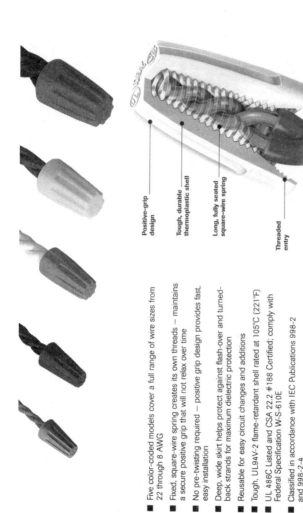

- Five color-coded models cover a full range of wire sizes from 22 through 8 AWG

- Fixed, square-wire spring creates its own threads – maintains a secure positive grip that will not relax over time

- No pre-twisting required – positive grip design provides fast, easy installation

- Deep, wide skirt helps protect against flash-over and turned-back strands for maximum dielectric protection

- Reusable for easy circuit changes and additions

- Tough, UL94V-2 flame-retardant shell rated at 105°C (221°F)

- UL 486C Listed and CSA 22.2 #188 Certified; comply with Federal Specification W-S-610E

- Classified in accordance with IEC Publications 998-2 and 998-2-4

FIGURE 10.15a Wire-Nut® wire connectors. (*Courtesy Ideal Industries*)

Positive-grip design

Tough, durable thermoplastic shell

Long, fully seated square-wire spring

Threaded entry

Deep skirt, wide throat

Model	Color	Wire Combination Range	Wire Combination Range (mm)	Quantity	Cat. No.
71B®	Gray	300V No. 22 to 16 AWG Min. 2 No. 22 – Max. 2 No. 16	300V .64mm DIA to 1.29mm DIA Min. .65mm² – Max. 2.62mm²	Box of 100 Carton of 1,000 Keg of 25,000 (loose pack)	30-071 30-171 30-271
72B®	Blue	300V No. 22 to 14 AWG Min. 2 No. 22 – Max. 3 No. 16	300V .64mm DIA to 1.63mm DIA Min. .65mm² Max. 3.93mm²	Box of 100 Carton of 1,000 Keg of 10,000 (loose pack)	30-072 30-172 30-272
73B®	Orange	600V* No. 22 to 14 AWG Min. 1 No. 18 & 1 No. 20 Max. 4 No. 16 & 1 No. 20	600V* .64mm DIA to 1.63mm DIA Min. 1.34mm² Max. 5.76mm²	Box of 100 Jar of 300 Carton of 1,000 Keg of 10,000 (20 bags, 500 ea.) Keg of 10,000 (loose pack)	30-073 30-073J 30-173 30-273 30-673
74B®	Yellow	600V* No. 18 to 12 AWG Min. 2 No. 18 Max. 4 No. 14 & 1 No. 18	600V* 1.02mm DIA to 2.05mm DIA Min. 1.64mm² Max. 9.14mm²	Box of 100 Jar of 175 Carton of 1,000 Keg of 10,000 (20 bags, 500 ea.) Keg of 10,000 (loose pack)	30-074 30-074J 30-174 30-274 30-674
76B®	Red	600V* No. 18 to 10 AWG Min. 2 No. 14 Max. 2 No. 10 & 2 No. 12	600V* 1.02mm DIA to 2.59mm DIA Min. 4.16mm² Max. 17.14mm²	Box of 100 Carton of 1,000 Keg of 5,000 (20 bags, 250 ea.)	30-076 30-176 30-276

*1,000V maximum in fixtures and signs

WIRE CONN
UL LISTED · CU/Cu Only · ℅ · CE · EN 60-998-2-4

FIGURE 10.15b Wire-Nut® wire connector, wire size combinations. (*Courtesy Ideal Industries*)

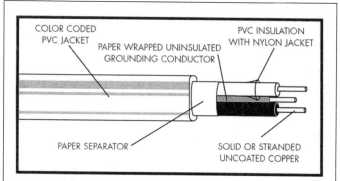

COLOR CODED PVC JACKET
PAPER WRAPPED UNINSULATED GROUNDING CONDUCTOR
PVC INSULATION WITH NYLON JACKET
PAPER SEPARATOR
SOLID OR STRANDED UNCOATED COPPER

Description & Features:

ESSEX® Type NM-B Non-metallic sheathed cable is available in both flat (2 conductors) and round (3 and 4 conductors) constructions. Flat constructions provide solid or stranded Type THHN conductors laid parallel with or without a paper wrapped uninsulated grounding conductor. Round constructions provide solid or stranded Type THHN conductors twisted together with or without a paper wrapped uninsulated grounding conductor and without fillers. Both flat, wrapped with a paper separator, and round assemblies are covered with a tough outer jacket of polyvinyl chloride (PVC).

- UL listed as Type NM-B E10816
- Conductors rated 90°C
- The ampacity shall be that of 60°C conductors and comply with NEC section 300.15
- Outer PVC jacket is color coded by conductor gauge size

#14 AWG – White
#12 AWG – Yellow
#10 AWG – Orange
#8 AWG and larger – Black

FIGURE 10.16a Type NM-B color coded non-metallic sheathed cable description. *(Courtesy Superior Essex)*

Applications:

■ Acceptable for installation in both exposed and concealed work in normally dry locations

■ Can be installed or fished in air voids, in masonry block or tile walls where such walls are not exposed or subjected to excessive moisture or dampness

■ Acceptable for use in accordance with NEC Articles 300.4(B) and 334

■ For use in residential & commercial wiring of branch circuits

Construction:

Conductors: Solid (#14, 12, & 10 AWG) (per ASTM B-3) or stranded (#8, 6, 4, & 2) soft annealed uncoated copper (per ASTM B-8)
Insulation: Polyvinyl chloride (PVC) compound to the thickness required by Underwriters Laboratories for Type THHN rated 90°C dry
COLOR CODE:
2 CONDUCTORS: Black, White
3 CONDUCTORS: Black, White, Red
4 CONDUCTORS **STYLE 1:** Black, White, Red, Blue
4 CONDUCTORS **STYLE 2:** Black, White, Red, White with Red Stripe
Conductor Jacket: Heat stabilized nylon
Grounding Conductor: Soft annealed solid or stranded uncoated copper
Separator: Paper (2 conductors only)
Assembly:
2 CONDUCTORS: Individual Type THHN conductors are laid parallel with or without a paper wrapped uninsulated grounding conductor placed between both conductors
3 and 4 CONDUCTORS: Individual Type THHN conductors are twisted together with or without a paper wrapped uninsulated grounding conductor and without fillers
Jacket: Polyvinyl chloride (PVC) applied over the assembly, high visibility color accent
Temperature: 60°C dry
Rating: 600 Volts

Specifications & Standards:

UL Standard 719 - Non-Metallic Sheathed Cable
UL Standard 83 - Insulated Conductors

*Conductors rated 90°C, outer jacket rated 60°C dry

FIGURE 10.16b Type NM-B color coded non-metallic sheathed cable applications. *(Courtesy Superior Essex)*

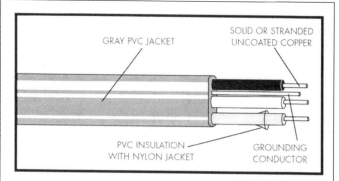

Description & Features:

ESSEX® Type UF-B Underground Feeder Cable's construction provides solid or stranded Type THHN/THWN conductors laid parallel with or without an uninsulated grounding conductor. An outer jacket of tough polyvinyl chloride (PVC), applied directly over the assembled conductors, provides protection against mechanical damage, moisture, chemicals and corrosion.

- UL listed as Type UF-B for direct burial E25682
- Grease and chemical resistant conductors
- Crush resistant
- Moisture, chemical and sunlight resistant jacket
- Conductors rated 90°C
- The ampacity shall be that of 60°C conductors and comply with NEC section 300.15

Sample Print:
ESSEX 12-2 WITH GROUND TYPE UF-B
SUNLIGHT RESISTANT 600V E25682* (UL)

FIGURE 10.17a Type UF-B underground feeder cable description. *(Courtesy Superior Essex)*

Applications:

■ Permitted for use underground including direct burial in earth as Feeder or Branch Circuit Cable where provided with overcurrent protection

■ For use in interior wiring in wet, dry or corrosive locations under the recognized wiring methods of the NEC

■ Acceptable to be installed as a Non-Metallic Sheathed Cable in accordance with NEC Article 340.10(4)

■ Acceptable for use in accordance with NEC Articles 300.4(B) and 340

Construction:

Conductors: Solid (per ASTM B-3) or stranded soft annealed uncoated copper (per ASTM B-8)
Insulation: Polyvinyl chloride (PVC) compound to the thickness required by Underwriters Laboratories for Type THHN 90°C rated
   COLOR CODE:
   2 CONDUCTORS: Black, White
   3 CONDUCTORS: Black, White, Red
Conductor Jacket: Heat stabilized nylon
Grounding Conductor: Uninsulated, soft annealed solid uncoated copper
Assembly: Individual Type THHN/THWN conductors are laid parallel with (or without) a grounding conductor. When supplied with a grounding conductor it is placed in the web between insulated conductors
Jacket: Gray sunlight resistant polyvinyl chloride (PVC) applied directly over the assembled conductors
Temperature: 60°C wet or dry
Rating: 600 Volts

Specifications & Standards:

UL Standard 493 - Thermoplastic-Insulated Underground Feeder and Branch Circuit Cables
UL Standard 83 - Thermoplastic-Insulated Wire and Cable (For the Insulated Conductors)

*Conductors rated 90°C, outer jacket rated 60°C dry

FIGURE 10.17b Type UF-B underground feeder cable applications. *(Courtesy Superior Essex)*

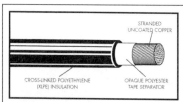

STRANDED UNCOATED COPPER

CROSS-LINKED POLYETHYLENE (XLPE) INSULATION

OPAQUE POLYESTER TAPE SEPARATOR

Description & Features:

ESSEX® Type USE-2/RHH/RHW-2 building wire is rated 600 volts for general purpose, new work or rewiring applications. Construction provides soft annealed stranded uncoated copper insulated with a cross-linked polyethylene (XLPE) compound that provides excellent abrasion, crush, chemical, and oil resistance.

- UL listed Type USE-2/RHH/RHW-2 90°C wet or dry E11134
- Resistant to acids, alkalis, grease and chemicals
- Abrasion and crush resistant
- Moisture, ozone resistant
- Resistant to sunlight (black only)

Sample Print:
ESSEX 1 AWG TYPE USE-2 OR TYPE RHH OR TYPE RHW-2 XLP 600V E11134* (UL)

Applications:

- Suitable for branch-circuit wiring, as a single conductor 600 volt building wire in accordance with the National Electrical Code where the maximum operating temperature does not exceed 90°C in wet or dry locations
- Suitable for direct burial or use in conduit and raceways installed underground, in wet locations and where condensation and moisture accumulations within the raceway may occur — National Electric Code's Article 338 "Service Entrance Cable"
- For use in accordance with the National Electric Code's Article 310 "Conductors for General Wiring" and Article 210 "Branch-Circuits"

Construction:

Conductor: Stranded soft annealed uncoated copper (per ASTM B-8)
Separator: Opaque polyester tape, as necessary
Insulation: Cross-linked polyethylene (XLPE) rated at 90°C wet or dry
Temperature: 90°C wet or dry
Rating: 600 Volts

Specifications & Standards:

UL Standard 44 - Rubber Insulated Wire and Cable
UL Standard 854 - Service Entrance Cables

FIGURE 10.18 Type USE-2/Rhh/RhW-2 single conductor description and application. *(Courtesy Superior Essex)*

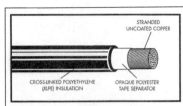

Description & Features:

ESSEX® Type XHHW-2 building wire is rated 600 volts for general purpose, new work or rewiring applications. Construction provides soft annealed stranded uncoated copper insulated with a cross-linked polyethylene (XLPE) compound that provides excellent abrasion, chemical, and oil resistance.

■ UL listed Type XHHW-2 90°C wet or dry E1139
■ Resistant to acids, alkalis, grease and chemicals
■ Abrasion resistant
■ Moisture, ozone resistant

Sample Print:
ESSEX 1 AWG TYPE XHHW-2 600VOLT E1139* (UL)

Applications:

■ Suitable for branch-circuit wiring, as a single conductor 600 volt building wire in accordance with the National Electrical Code where the maximum operating temperature does not exceed 90°C in wet or dry locations

■ Suitable for use in conduit and raceways installed underground in conduit or raceway, in wet locations and where condensation and moisture accumulations within the raceway may occur

■ For use in accordance with the National Electric Code's Article 310 "Conductors for General Wiring" and Article 210 "Branch-Circuits"

Construction:

Conductor: Stranded soft annealed uncoated copper (per ASTM B-8) or UL 44
Separator: Opaque polyester tape, as necessary
Insulation: Cross-linked polyethylene (XLPE) rated at 90°C wet or dry
Temperature: 90°C wet or dry
Rating: 600 Volts

Specifications & Standards:

UL Standard 44 - Rubber Insulated Wire and Cable

FIGURE 10.19 Type XHHW-2 single conductor description and application. *(Courtesy Superior Essex)*

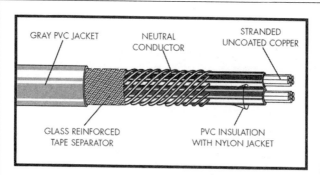

Description & Features:

ESSEX® Type SE-U Service Entrance Cable's construction provides stranded Type THHN/THWN conductors laid parallel and helically wrapped with an uninsulated solid neutral conductor. A glass reinforced tape is applied over the assembled conductors and covered with a jacket of tough polyvinyl chloride (PVC) that provides protection against mechanical damage, moisture, chemicals and corrosion.

- UL listed as Type SE, E-11134
- Alkali, grease and chemical resistant conductors
- Oil and gasoline resistant conductors
- Moisture, chemical and corrosion resistant jacket
- Crush resistant jacket
- Resistant to sunlight

Sample Print:
ESSEX 6-6-6 TYPE SE CABLE TYPE
THHN/THWN CDRS. 600V E11134* (UL)

FIGURE 10.20a Type SE-U service entrance cable description. *(Courtesy Superior Essex)*

Applications:

■ Acceptable for use as Service Drop Cable from secondary wires at the pole to point of attachment to the building (required by NEC Article 230) may be used inside premises for branch circuit (specified by NEC Section 338.11).

■ Acceptable for use as Service Entrance Cable between the terminals of the service equipment and the point of connection to the service drop or service lateral

■ Acceptable for use as a combination service drop and service entrance, permitting a continuous, unspliced connection between the secondary wires at the pole and the service equipment

■ Acceptable for use in accordance with NEC Article 338 - Service Entrance Cable

Construction:

Conductors: Stranded soft annealed uncoated copper (per ASTM B-8) or UL83
Insulation: Polyvinyl chloride (PVC) compound to the thickness required by Underwriters Laboratories for 90°C rated Type THHN
Conductor Jacket: Heat stabilized nylon
Neutral Conductor: Soft annealed solid uncoated copper
Separator: Glass reinforced tape
Assembly: Individual Type THHN/THWN conductors are laid parallel (without filler) with the neutral conductor applied helically around the assembled conductors and covered with a suitable glass reinforced tape
Jacket: Gray polyvinyl chloride (PVC)
Temperature: 90°C dry and 75°C wet
Rating: 600 Volts

Specifications & Standards:
UL Standard 854 - Service Entrance Cables

FIGURE 10.20b Type SE-U service entrance cable applications. *(Courtesy Superior Essex)*

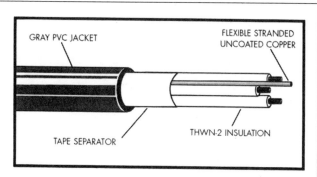

Description & Features:

ESSEX® Type SER Service Entrance Cable's construction provides stranded Type THHN/THWN-2 conductors & neutral or stranded Type THHN/THWN-2 conductors & neutral & a bare equipment ground. A glass reinforced tape is applied over the assembled conductors and covered with a jacket of tough polyvinyl chloride (PVC) that provides protection against mechanical damage, moisture, chemicals and corrosion.

- UL listed as Type SE, E-11134
- Alkali, grease and chemical resistant conductors
- Oil and gasoline resistant conductors
- Moisture, chemical and corrosion resistant jacket
- Crush resistant jacket
- Resistant to sunlight

Sample Print:
ESSEX 6-6-6 TYPE SE STYLE R THHN/THWN-2 CDRS. 600V E11134* (UL)

FIGURE 10.21a Type SER service entrance cable description. *(Courtesy Superior Essex)*

Applications:

- Typically used as a panel feeder in multiple unit dwellings

- Acceptable for use as Service Drop Cable from secondary wires at the pole to point of attachment to the building (required by NEC Article 230) may be used inside premises for branch circuit (specified by NEC Section 338.11).

- Acceptable for use as Service Entrance Cable between the terminals of the service equipment and the point of connection to the service drop or service lateral

- Acceptable for use as a combination service drop and service entrance, permitting a continuous, unspliced connection between the secondary wires at the pole and the service equipment

- Acceptable for use in accordance with NEC Article 338 - Service Entrance Cable

Construction:

Conductors: Stranded soft annealed uncoated copper (per ASTM B-8) or UL83
Insulation: Polyvinyl chloride (PVC) compound to the thickness required by Underwriters Laboratories for 90°C rated Type THHN
Conductor Jacket: Heat stabilized nylon
Neutral Conductor: Soft annealed solid uncoated copper
Separator: Glass reinforced tape
Assembly: Type THHN/THWN-2 conductors & neutral & a bare equipment ground
Jacket: Gray polyvinyl chloride (PVC)
Temperature: 90°C dry and 90°C wet
Rating: 600 Volts

Specifications & Standards:

UL Standard 854 - Service Entrance Cables
UL Standard 83 - Thermoplastic Insulated Wires & Cables

FIGURE 10.21b Type SER service entrance cable applications. *(Courtesy Superior Essex)*

Applications:

- Used in overhead transmission and distribution systems
- Used for grounding grid systems

Construction:

Conductor:

SOFT DRAWN (Temper):
Solid ASTM B3...............14 AWG - 2 AWG
Stranded ASTM B8 or B787....8 AWG - 500 MCM

Specifications & Standards:

ASTM B3 - Soft or Annealed Copper Wire
ASTM B8 - Concentric-Lay-Stranded Copper
  Conductors
ASTM, B787 Combination Unilay Strand
  Conductors

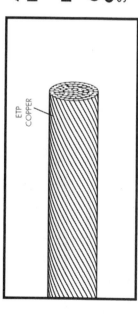

ETP COPPER

Description & Features:

ESSEX® solid and stranded bare copper conductors are available (soft annealed copper) and sizes (14 AWG to 500 MCM). Electrolytic Tough Pitch (ETP) copper feed stock is used to draw solid conductors (and B3). Solid and stranded conductors (per ASTM B8 or ASTM B787) are available in stock or customer specified packaging.

- Highest conductivity per unit area of all commonly used metals to conduct electricity
- Flexible, easily shaped and formed into place
- Easily welded and soldered

FIGURE 10.22a Bare copper wire—solid and stranded description. (*Courtesy Superior Essex*)

BARE COPPER WIRE/Solid & Stranded, Soft

P/N	Size AWG or kcmil	No. of Strands	O.D. (Inches)	Standard Package	Product Wgt. Lbs./Mft.
060142C*	14	SOLID	.064	H	12.4
060122C*	12	SOLID	.081	G	19.8
060102C*	10	SOLID	.102	E	31.4
060082C*	8	SOLID	.128	D,F	50
060062C*	6	SOLID	.162	C,F	79
060042C*	4	SOLID	.204	B,F	126
060022C*	2	SOLID	.258	A	201
060081C*	8	7	.148	D,J	51
060061C*	6	7	.186	C,F	81
060041C*	4	7	.234	B,F	129
060021C*	2	7	.269	A,F	205
060011C*	1	19	.321	D	259
061101C*	1/0	19	.374	F	326
062101C*	2/0	19	.419	D,F	411
063101C*	3/0	19	.470	F	518
064101C*	4/0	19	.528	D,F,J	653
062501C800	250	37	.575	J	772
063501C800	350	37	.681	J	1080
065001C800	500	37	.813	J	1544

*Consult your ESSEX sales representative for package codes and availability.

All diameters are nominal values; all weights are product weights exclusive of packaging.
All diameters and weights are subject to normal manufacturing tolerances.

NR (Non-returnable Reel)

NOTE: Medium Hard Drawn and Hard Drawn wires
are available subject to minimum order quantities.

FIGURE 10.22b Bare copper wire—solid and stranded application. *(Courtesy Superior Essex)*

220

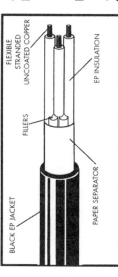

FLEXIBLE
STRANDED
UNCOATED COPPER

EP INSULATION

FILLERS

BLACK EP JACKET

PAPER SEPARATOR

Applications:

■ General purpose heavy-duty service for garages, portable lights, battery chargers, portable stage lights, heavy tools and equipment

■ For use in accordance with the NEC Article 400

Construction:

Conductors: Extra flexible, Class K stranded soft annealed uncoated copper

Insulation: EP Rubber

Assembly: Conductors are cabled with fillers (as necessary) and a suitable paper separator

Jacket: Black thermoset EP rubber

Temperature: 75°C UL/ 60°C CSA

Voltage: 300 Volts (Type SJ)

Description & Features:

Essex®/Royal® Type SJ general purpose, heavy-duty, flexible cord provides flexible Class K stranded conductors insulated with ethylene propylene (EP) rubber. The conductors are cabled with fillers and jacketed with a tough EP rubber compound.

■ UL Listed Type SJ (300V) 75°C
■ CSA Certified Type SJ (300V) 60°C
■ Meets UL (Horizontal Specimen) and CSA (FT2) Flame Test Requirements

Specifications & Standards:

UL Standard 62
CSA Standard C22.2 No.49

MADE IN
THE USA

UL Listed

Certified
Canadian Standards Association

FIGURE 10.23a Type SJ flexible cord description. *(Courtesy Superior Essex)*

| Catalog No. | Conductor | | | Nominals (INCHES) | | | Stock Lengths | | Ampacity§ NEC | Product Wgt. Lbs./Mft. |
	No.	AWG	Strand	Insul.	Jkt.	O.D.	Standard Put-Up*	Other Available Lengths**		
431821*	2	18	16 x 30	.030	.030	.29	250' Spl, 1000' RI	-	10	45
431831*	3	18	16 x 30	.030	.030	.31	250' Spl, 1000' RI	-	10	55
431841*	4	18	16 x 30	.030	.030	.34	-	250' Spl, 1000' RI	7	70
431621*	2	16	26 x 30	.030	.030	.31	250' Spl	1000' RI	13	55
431631*	3	16	26 x 30	.030	.030	.33	250' Spl, 1000' RI	-	13	70
431641*	4	16	26 x 30	.030	.030	.37	250' Spl	1000' RI	10	85
431421*	2	14	41 x 30	.030	.030	.34	250' Spl	1000' RI	18	75
431431*	3	14	41 x 30	.030	.030	.37	250' Spl, 1000' RI	-	18	95
431441*	4	14	41 x 30	.030	.030	.41	250' Spl	1000' RI	15	120
431221*	2	12	65 x 30	.030	.045	.42	250' Spl, 1000' RI	250' Spl, 1000' RI	25	110
431231*	3	12	65 x 30	.030	.045	.44	250' Spl, 1000' RI	-	25	145
431241*	4	12	65 x 30	.030	.045	.48	250' RI	1000' RI	20	180
431021*	2	10	104 x 30	.045	.060	.57	-	250' RI, 1000' RI	30	170
431031*	3	10	104 x 30	.045	.060	.60	250' RI	1000' RI	30	235
431041*	4	10	104 x 30	.045	.060	.66	-	250' RI, 1000' RI	25	290

§Per Table 400.5 (A) of the NEC

**Consult your Essex/Royal Sales Representative for minimum production runs.

*Consult your Essex/Royal Sales Representative for cable color code and packaging code digits.
All diameters are nominal values; all weights are product weights exclusive of packaging.
All diameters and weights are subject to normal manufacturing tolerances.

COLOR CODE:
2/C BLACK-WHITE
3/C BLACK-WHITE-GREEN
4/C BLACK-WHITE-RED-GREEN

FIGURE 10.23b   Type SJ flexible cord application. (Courtesy Superior Essex)

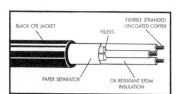

BLACK CPE JACKET

FLEXIBLE STRANDED UNCOATED COPPER

FILLERS

PAPER SEPARATOR

OIL RESISTANT EPDM INSULATION

Applications:

■ Recommended for heavy-duty service where cord is subject to indoor and outdoor use
■ For garages, heavy tools, portable power equipment and equipment exposed to water, oils, acids and chemicals
■ Portable lights, battery chargers and portable stage lights
■ Suitable for use in wet environments including submersion in water
■ For use in accordance with the NEC Article 400

Description & Features:

ESSEX®/ROYAL® Type SOOW & Type SJOOW are the most popular heavy-duty indoor and outdoor, general purpose, service cords. Construction provides flexible Class K stranding and water and oil resistant ethylene propylene (EP) rubber insulation. Combined with a tough, proven chlorinated polyethylene (CPE) jacketing compound this cord provides extended service life and water, oil and chemical resistance in indoor and outdoor industrial uses.

■ UL listed & CSA certified SOOW/600V
■ UL listed & CSA certified SJOOW/300V
■ Water resistant insulation
■ Oil resistant insulation
■ Weather, oil, acid, ozone and sunlight resistant jacket
■ Abrasion resistant
■ UL listed and CSA certified as weather resistant for outdoor applications
■ Meets MSHA, UL (Horizontal Specimen) and CSA (FT2) flame test requirements

Construction:

Conductors: Extra flexible, Class K stranded soft annealed uncoated copper
Insulation: Water and Oil Resistant EP Rubber
Assembly: Conductors are cabled together with fillers (as necessary) and a suitable tape separator
Jacket: Black thermoset chlorinated polyethylene (CPE) rubber
Temperature: 90°C to -40°C, Water and Oil Resistant 60°C
Voltage: 600 Volts (Type SOOW)
   300 Volts (Type SJOOW)

Specifications & Standards:

UL Standard 62
CSA Standard C22.2 No. 49
MSHA and Pennsylvania DEP approved and marked

UL LISTED

Certified Canadian Standards Association

MADE IN THE USA

FIGURE 10.24 Type SOOW and type SJOOW flexible cord description and application. *(Courtesy Superior Essex)*

Applications:

- Suitable for use as branch wiring drops from busways for connection to portable or stationary equipment
- For use in accordance with the NEC Article 368

Construction:

Conductors: Soft annealed stranded uncoated copper

Insulation: Polyvinyl chloride (PVC) compound, rated 60°C, meeting UL requirements for Type TW wire. Color coded black, white and red

Grounding Conductor(s): Uninsulated soft annealed stranded uncoated copper

Assembly: Conductors are cabled together with uninsulated grounding conductor(s) filler as required and a suitable tape separator.

Jacket: Gray polyvinyl chloride (PVC)

Temperature: 60°C

Rating: 600 volts

Specifications & Standards:

UL Standard 83 - Thermoplastic insulated wires and cables

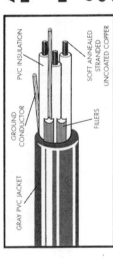

GRAY PVC JACKET

GROUND CONDUCTOR

PVC INSULATION

SOFT ANNEALED STRANDED UNCOATED COPPER

FILLERS

Description & Features:

ESSEX®/ROYAL® UL Listed Bus Drop Cable construction provides stranded soft annealed uncoated copper conductors insulated with polyvinyl chloride (PVC) and cabled together with uninsulated grounding conductor(s) and fillers as required. An outer jacket of a tough PVC compound provides excellent resistance to oils, grease lubricants, gasoline, acids, chemicals and cutting fluids.

- UL listed Bus Drop Cable
- Resistant to oils, grease lubricants, gasoline, acids, chemicals and cutting fluids

MADE IN THE USA

(UL) LISTED

FIGURE 10.25a Bus drop cable description. (*Courtesy Superior Essex*)

BUS DROP CABLE 60°C/600V

Cat. No.	AWG Size	Ground No. x AWG	Nominals (Inches)			Standard Put-Up & Shipping Package	Ampacity NEC §	Product Wgt. (LBS./MFt.)
			Insul.	Jacket	O.D.			
191231G*	12/3	1 x 12	.030	.045	.44	250' Ctn, 1000 Reel	20	150
191031G*	10/3	1 x 10	.030	.045	.50	250' Ctn, 1000 Reel	30	210
190831G*	8/3	1 x 10	.045	.060	.66	250' Reel	40	340
190631G*250	6/3	1 x 10	.060	.060	.81	250' Reel	55	495
190431G*250	4/3	1 x 8	.060	.080	.95	250' Reel	70	740
190231G*250	2/3	1 x 8	.060	.080	1.07	250' Reel	95	1035

§ Based on not more than three conductors in a cable in free air with an ambient temperature of 30°C and a conductor temperature of 60°C

* Consult your Essex/Royal sales representative for package codes and availability.
All diameters are nominal values; all weights are product weights exclusive of packaging. All diameters and weights are subject to normal manufacturing tolerances.

FIGURE 10.25b Bus drop cable application. (*Courtesy Superior Essex*)

Applications:

- For use in power supply cords for household fans, clocks, lamps, radios, and small display signs and similar appliances and general use extension cords where the cord is not subjected to hard use
- For use in accordance with the NEC Article 400

Construction:

Conductors: Soft annealed flexible stranded uncoated copper per ASTM B-174

Assembly: Two conductors are laid flat and parallel

Insulation/Jacket: Polyvinyl chloride (PVC) applied directly over the conductors

Temperature: 60°C continuous in air

Voltage: 300 Volts

Specifications & Standards:

UL Standard 62 - Flexible Cord and Fixture Wire
CSA Standard C22.2 No. 49

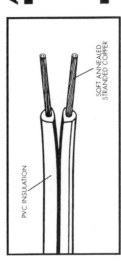

PVC INSULATION

SOFT ANNEALED
STRANDED COPPER

Description & Features:

ESSEX®/ROYAL® UL listed and CSA certified Type SPT Parallel Cord is a general purpose, light-duty flexible cord. Construction provides two flexible stranded conductors laid flat and parallel and covered with an integral insulation and jacket of polyvinyl chloride (PVC). A "mid rip" between conductors provides for easy splitting and stripping.

- UL listed SPT 1,2 Parallel Cord (300 Volts)
- CSA Certified SPT 1,2 Parallel Cord (300 Volts)
- Class M extra flexible stranding for 18 AWG
- Rated 60°C continuous in air
- Meets CSA FT2 flame test requirements

FIGURE 10.26a Type SPT parallel cord description. (*Courtesy Superior Essex*)

TYPE SPT-1 PARALLEL CORD 60°C/300V

Cat. No.	Conductor AWG	Conductor Strand	No.	Nominals (Inches) Insul.	Nominals (Inches) O.D.	Colors	Standard Lengths Put-Up	Standard Lengths Ship. Pkg.	Ampacity §	Product Wgt. (LBS./MFt.)
5916210252	16 SPT-2	26 × 30	2	.048	.156 × .307	BLACK	250' Spool	1000'	13	36
5916211252	16 SPT-2	26 × 30	2	.048	.156 × .307	WHITE	250' Spool	1000'	13	36
5916217252	16 SPT-2	26 × 30	2	.048	.156 × .307	BROWN	250' Spool	1000'	13	36

Cat. No.	Conductor AWG	Conductor Strand	No.	Nominals (Inches) Insul.	Nominals (Inches) O.D.	Colors	Standard Lengths Put-Up	Standard Lengths Ship. Pkg.	Ampacity §	Product Wgt. (LBS./MFt.)
5918210252	18 SPT-1	41 X 34	2	.033	.113 × .250	BLACK	250' Spool	1000'	10	20.9
5918211252	18 SPT-1	41 X 34	2	.033	.113 × .250	WHITE	250' Spool	1000'	10	20.9
5918217252	18 SPT-1	41 X 34	2	.033	.113 × .250	BROWN	250' Spool	1000'	10	20.9

§ Per table 400.5 (A) of the NEC
† Extension cords only. Specify 65 x 34 for Power Supply Cords.

All diameters are nominal values; all weights are product weights exclusive of packaging. All diameters and weights are subject to normal manufacturing tolerances.

FIGURE 10.26b Type SPT parallel cord application. (*Courtesy Superior Essex*)

Applications:

■ Thermostat Cable

■ Designed for applications where code requirements and/or specifications require a ETL Listed CL2 Power Limited Circuit Cable

■ Used for air conditioning controls, heat pumps, heating controls, thermostats and bell and alarm systems

■ For use in accordance with the NEC Article 725

Construction:

Conductors: Soft annealed solid uncoated copper
Insulation: Colored coded polyvinyl chloride (PVC)
Assembly: Conductors are twisted or laid parallel without fillers
Jacket: Polyvinyl chloride (PVC)applied directly over the assembled conductors, 20 AWG brown, 18 AWG white
Temperature: 60°C continuous in air
Voltage: None (see NEC article 725) (30 volts)

Specifications & Standards:

Conforms to UL Std 13

PVC JACKET

PVC INSULATION

SOLID ANNEALED
UNCOATED COPPER

Description & Features:

ESSEX® ETL listed Type CL2 Power Limited Circuit Cable is a general purpose multiconductor low voltage control cable. Construction provides solid uncoated copper conductors insulated with a polyvinyl chloride (PVC) insulation twisted together without fillers and covered with a ETL listed sunlight resistant PVC jacket.

■ ETL listed Type CL2 Power Limited Circuit Cable
■ ETL listed sunlight resistant jacket
■ Rated 60°C continuous in air
■ Meets UL 1581 vertical flame tray test

Insulation Color Coded Polyvinyl Chloride PVC
18 AWG white jacket, 20 AWG brown jacket
2 = Red, white
3 = Red, white, green
4 = Red, white, green, blue
5 = Red, white, green, blue, yellow
6 = Red, white, green, blue, yellow, brown
7 = Red, white, green, blue, yellow, brown, orange
8 = Red, white, green, blue, yellow, brown, orange, black
9 = Red, white, green, blue, yellow, brown, orange, black, pink
10 = Red, white, green, blue, yellow, brown, orange, black, pink and tan

FIGURE 10.27a Type CL2 thermostat cable description. (*Courtesy Superior Essex*)

TYPE CL2 - POWER LIMITED CIRCUIT CABLE 60°C

Cat. No.	Conductor AWG	Conductor Strand	Conductor No.	Nominal (Inches) Insulation	Nominal (Inches) Jacket	Nominal (Inches) O.D.	Standard Length (ft.) Put-Up	Standard Length (ft.) Shp. Pkg.	Product Wgt. (LBS./MFt.)
162022 7502	20	Solid	2	.006	.015	.070 x .112	500' Spool	2000'	11
162032 7502	20	Solid	3	.006	.015	.124	500' Spool	2000'	15
162042 7252	20	Solid	4	.006	.015	.135	250' Spool	1000'	20
162052 7252	20	Solid	5	.006	.015	.148	250' Spool	1000'	24
162062 7252	20	Solid	6	.006	.015	.156	250' Spool	1000'	27
162072 7252	20	Solid	7	.006	.015	.161	250' Spool	1000'	31
162082 7252	20	Solid	8	.006	.015	.175	250' Spool	1000'	35
160202 72520N	20	Solid	10	.006	.015	.191	250' Spool	1000'	43
161822 1502	18	Solid	2	.006	.015	.091 x .135	500' Spool	2000'	15
161832 1502	18	Solid	3	.006	.015	.144	500' Spool	2000'	21
161842 1252	18	Solid	4	.006	.015	.158	250' Spool	1000'	27
161852 1252	18	Solid	5	.006	.015	.174	250' Spool	1000'	34
161862 1252	18	Solid	6	.006	.015	.184	250' Spool	1000'	40
161872 1252	18	Solid	7	.006	.015	.190	250' Spool	1000'	45
161882 1252	18	Solid	8	.006	.015	.208	250' Spool	1000'	51
160182 12520N	18	Solid	10	.006	.015	.227	250' Spool	1000'	63

All diameters are nominal values; all weights are product weights exclusive of packaging. All diameters and weights are subject to normal manufacturing tolerances.
* Consult your ESSEX sales representative for package codes and availability.

FIGURE 10.27b Type CL2 thermostat cable application. (*Courtesy Superior Essex*)

Applications:

■ Fire protection systems

■ Designed for applications where code requirements and/or specifications require an FPLR or FPLP cable

Construction:

Conductors: Soft drawn bare copper
Insulation: PVC 2000 PSI minimum
Assembly: Conductors are twisted or laid parallel without fillers
Jacket: Red PVC
Temperature: -20°C to 60°C
Voltage: See NEC Article 760

Specifications & Standards:

Conforms to UL Std 1424, N.E.C. Article 760, CMR or CMP Rated

MADE IN THE USA

RED PVC JACKET

* PVC INSULATION

SOLID ANNEALED UNCOATED COPPER

Description & Features:

ESSEX® ETL listed unshielded fire alarm cable is used where fire protective signalling circuitry is required. Available in both CMR (riser) and CMP (plenum) rated.

■ ETL listed unshielded fire alarm cable, Type FPLR or Type FPLP.
■ Red jacket.
■ Rated 60°C continuous in air

Insulation Color Coded Polyvinyl Chloride PVC

2 – Red, black
4 – Red, black, green, white
6 – Red, black, green, white, brown, blue
8 – Red, black, green, white, brown, blue, orange, yellow

FIGURE 10.28a Type FPLR/FPLP fire alarm cable description. (*Courtesy Superior Essex*)

TYPE FPLR

ITEM #	PRODUCT AWG/COND	SHIPPING LBS/MFT	COPPER LBS/MFT	NOM. WALL THICKNESS		NOMINAL O.D.
				INSULATION	JACKET	
T112222602	12/2	56	39.7	.013	.018	.250
T114222602	14/2	36	24.8	.013	.018	.218
T114422602	14/4	65	49.7	.013	.018	.255
T116222602	16/2	20	15.6	.013	.018	.174
T116422602	16/4	37	31.2	.013	.018	.205
T118222602	18/2	17	10.1	.010	.015	.146
T118422602	18/4	25	20.2	.010	.015	.178
T118622602	18/6	45	31.5	.010	.015	.208
T118822602	18/8	60	42.5	.010	.015	.238

TYPE FPLP

ITEM #	PRODUCT AWG/COND	SHIPPING LBS/MFT	COPPER LBS/MFT	NOM. WALL THICKNESS		NOMINAL O.D.
				INSULATION	JACKET	
T312222602	12/2	48	39.7	.012	.014	.222
T314222602	14/2	32	24.8	.012	.014	.190
T314422602	14/4	62	49.7	.012	.014	.223
T316222602	16/2	21	15.6	.009	.014	.156
T316422602	16/4	38	31.2	.009	.014	.187
T318222602	18/2	14	10.1	.009	.014	.138
T318422602	18/4	23	20.2	.009	.014	.155
T318622602	18/6	38	30.3	.009	.014	.165
T318822602	18/8	54	40.4	.009	.014	.208

FIGURE 10.28b Type FPLR/FPLP fire alarm cable application. (Courtesy Superior Essex)

Applications:

■ **Fire protection systems**
■ Designed for applications where code requirements and/or specifications require an FPLR or FPLP cable

Construction:

Conductors: Soft drawn bare copper
Insulation: PVC 2000 PSI minimum
Assembly: Conductors are twisted or laid parallel without fillers
Jacket: Red PVC
Temperature: -20°C to 60°C
Voltage: See NEC Article 760

Specifications & Standards:

Conforms to UL Std 1424, NEC Article 760, CMR or CMP Rated

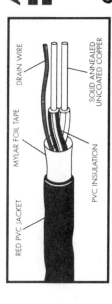

RED PVC JACKET
MYLAR FOIL TAPE
DRAIN WIRE
PVC INSULATION
SOLID ANNEALED UNCOATED COPPER

Description & Features:

ESSEX® ETL listed shielded fire alarm cable is used where fire protective signalling circuitry is required. Available in both CMR (riser) and CMP (plenum) rated.

■ ETL listed unshielded fire alarm cable, Type FPLR or Type FPLP.
■ Red jacket.
■ Rated 60°C continuous in air

Insulation Color Coded Polyvinyl Chloride PVC

2 – Red, black
4 – Red, black, green, white
6 – Red, black, green, white, brown, blue
8 – Red, black, green, white, brown, blue, orange, yellow

FIGURE 10.29a Type FPLR/FPLP shielded fire alarm cable description. (*Courtesy Superior Essex*)

232

TYPE FPLR SHIELDED

ITEM #	PRODUCT AWG/COND	SHIPPING LBS/MFT	COPPER LBS/MFT	NOM. WALL THICKNESS INSULATION	NOM. WALL THICKNESS JACKET	NOMINAL O.D.
T212222602	12/2	58	41.5	.013	.018	.262
T214222602	14/2	38	26.8	.013	.018	.230
T214422602	14/4	67	51.6	.013	.018	.269
T216222602	16/2	22	17.6	.013	.018	.191
T216622602	16/4	39	33.2	.013	.018	.223
T218222602	18/2	19	11.8	.010	.015	.170
T218422602	18/4	27	21.6	.010	.015	.197
T218622602	18/6	45	31.5	.010	.015	.244
T218822602	18/8	60	42.5	.010	.015	.280

TYPE FPLP SHIELDED

ITEM #	PRODUCT AWG/COND	SHIPPING LBS/MFT	COPPER LBS/MFT	NOM. WALL THICKNESS INSULATION	NOM. WALL THICKNESS JACKET	NOMINAL O.D.
T412222602	12/2	51	41.5	.012	.014	.234
T414222602	14/2	35	26.8	.012	.014	.202
T414422602	14/4	65	51.6	.012	.014	.235
T416222602	16/2	24	17.6	.009	.014	.168
T416422602	16/4	43	33.2	.009	.014	.199
T418222602	18/2	17	11.8	.009	.014	.150
T418422602	18/4	26	21.6	.009	.014	.167
T418622602	18/6	41	31.4	.009	.014	.177
T418822602	18/8	57	41.2	.009	.014	.220

FIGURE 10.29b Type FPLR/FPLP shielded fire alarm cable application. (*Courtesy Superior Essex*)

Applications:

■ Security cable

- Designed for applications where code requirements and/or specifications require a CL2r power limited circuit cable
- Used for security systems, intercoms, and speakers
- For use in accordance with NEC Article 725

Construction:

Conductors: Soft drawn bare copper
Insulation: PVC 2000 PSI minimum
Assembly: Conductors are twisted or laid parallel without fillers
Jacket: PVC 2500 PSI minimum
Shield: 100% aluminum/Mylar with 22 AWG tinned copper drain wire
Temperature: -20°C to 60°C
Voltage: See NEC article 725

Specifications & Standards:

Type CL2R, Conforms to UL 13, NEC Article 725, CMR (Riser) Rated

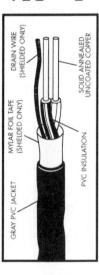

GRAY PVC JACKET
MYLAR FOIL TAPE (SHIELDED ONLY)
DRAIN WIRE (SHIELDED ONLY)
SOLID ANNEALED UNCOATED COPPER
PVC INSULATION

Description & Features:

ESSEX® ETL listed security cable can be used in a number of low voltage applications including security systems, intercoms, speakers, and other applications that require a CL2r power limited circuit cable.

- ETL listed security cable, Type CL2R, CMR
- Gray jacket
- Shielded or unshielded

Insulation Color Coded Polyvinyl Chloride PVC

2 – Red, black
4 – Red, black, green, white

MADE IN THE USA

FIGURE 10.30a Type CL2R security cable description. *(Courtesy Superior Essex)*

TYPE CL2R

ITEM #	PRODUCT AWG/COND	SHIPPING LBS/MFT	COPPER CONSTR.	NOM. WALL THICKNESS INSULATION	JACKET	NOMINAL O.D.
T52222G601	22/2	11	SOLID	.008	.016	.127
T52242G601	22/4	17	SOLID	.008	.016	.142
T52221G601	22/2	12	7/30	.008	.016	.130
T52241G601	22/4	18	7/30	.008	.016	.149
T51821G601	18/2	19	16/30	.008	.016	.160
T51841G601	18/4	33	16/30	.008	.016	.187
T51621G601	16/2	24	26/30	.008	.016	.186
T51641G601	16/4	43	26/30	.008	.016	.217
T51421G601	14/2	38	19/.0147	.008	.016	.230
T51441G601	14/4	69	19/.0147	.008	.016	.242

TYPE CL2R SHIELDED

ITEM #	PRODUCT AWG/COND	SHIPPING LBS/MFT	COPPER CONSTR.	NOM. WALL THICKNESS INSULATION	JACKET	NOMINAL O.D.
T62222G602	22/2	15	7/30	.008	.016	.134
T62242G602	22/4	21	7/30	.008	.016	.153
T61822G602	18/2	22	16/30	.008	.016	.164
T61842G602	18/4	36	16/30	.008	.016	.196
T61622G602	16/2	27	26/30	.008	.016	.195
T61642G602	16/4	46	26/30	.008	.016	.225
T61422G602	14/2	41	19/.0147	.008	.016	.237
T61442G602	14/4	72	19/.0147	.008	.016	.250

FIGURE 10.30b Type CL2R security cable application. *(Courtesy Superior Essex)*

Applications:

- Suitable for machine tool and appliance wire applications at temperatures from -25°C to 105°C
- Used for general purpose wiring in control cabinets
- For use in accordance with the National Electric Code's Article 310 and NFPA Standard 79

Construction:

Conductor: Soft annealed stranded uncoated copper

Insulation: High quality moisture resistant and VW-1 rated Polyvinyl chloride (PVC) compound

Temperature: 90°C MTW and AWM
105°C AWM and CSA TEW

Voltage:
1. 600 Volts AWM
2. 600 Volts MTW
3. 1000 Volts AWM
4. 600 Volts TEW

Specifications & Standards:

UL Standard 1063 - Machine Tool Wires & Cables
UL Standard 758 - Appliance Wire
CSA Standard C 22.2 No.127

 Certified
Canadian Standards Association

(UL) LISTED

MADE IN THE USA

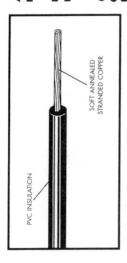

PVC INSULATION

SOFT ANNEALED
STRANDED COPPER

Description & Features:

ESSEX®/ROYAL® Quad-Rated® UL listed Machine Tool and Appliance Wire and CSA Certified TEW Wire is a superior multi-purpose product. Construction provides stranded soft annealed uncoated copper conductor insulated with a VW-1 rated polyvinyl chloride (PVC) compound that offers excellent abrasion, acid, chemical, oil and moisture resistance.

- UL Listed Machine Tool Wire (MTW) 90°C
- UL Recognized Appliance Wiring Material (AWM) 105°C 600 Volts and 90°C 1000 Volts (18 AWG to 10 AWG)
- CSA Certified as TEW 105°C
- Rated Oil Resistant at 60°C
- Meets UL VW-1 and CSA FT1 flame test requirements
- Resistant to acids, alkali, grease and chemicals
- Abrasion and moisture resistant

FIGURE 10.31a Machine tool and appliance wire description. *(Courtesy Superior Essex)*

QUAD-RATED® MACHINE TOOL & APPLIANCE WIRE 90°C/600V

CAT. NO.	AWG SIZE	INSULATION THICKNESS	STRANDING	NOMINAL O.D. (In.)	CU WGT. (Lbs./Mfr.)	STANDARD LENGTHS		PRODUCT WGT. (Lbs./Mfr.)
						PUT-UP	500' SPL SHIP. PKG.	
				STANDARD WALL				
240181*QR	18	.030	16 x 30 AWG	.11	4.92	A,B	2000'	10
240161*QR	16	.030	26 x 30 AWG	.12	7.99	A,B,E	2000'	13
240141*QR	14	.030	19 STR	.14	12.57	A,B,E	2000'	19
240121*QR	12	.030	19 STR	.16	19.96	A,B	2000'	28
240101*QR	10	.030	19 STR	.18	31.61	A,B	1000'	41
240081*QR	8	.045	19 STR	.24	50.02	A,D	500'	69
240061*HVQR	6	.060	19 STR	.32	80.98	C	500'	110
240041*HVQR	4	.060	19 STR	.36	128.78	C	500'	166
240021*HVQR	2	.060	19 STR	.42	204.48	C	500'	250

STANDARD PACKAGE CODE

A   500' Spool   C   500' Reel   E   5000' Reel
B   2500' Reel   D   1500' Reel

* Consult your ESSEX/ROYAL sales representative for package codes, availability, and/or made to order colors.
All diameters are nominal values; all weights are product weights exclusive of packaging. All diameters and weights are subject to normal manufacturing tolerances.

FIGURE 10.31b   Machine tool and appliance wire application. (*Courtesy Superior Essex*)

TOOLS

E lectricians require a number of tools to accomplish their work. Some of the tools are common among many trades, but some are specialized for the electrical trade. Having the right tools can make any job easier, faster, and more profitable. Selecting the right tools is not always easy. Individuals who don't have a lot of experience may buy the wrong tools. It takes years of experience to appreciate the need for special tools. Seasoned electricians know how important the type and quality of a tool is. Fortunately, you have this chapter to guide you along the path to proper tools.

LADDERS

Ladders are used frequently in the electrical trade. This is a tool where you don't want to scrimp on quality. You will need stepladders and extension ladders. A basic electrical truck should have a 6 foot stepladder, an 8 foot stepladder, and a 20 foot extension ladder. A 40 foot extension ladder may also be needed. Extension ladders should be equipped with rubber shoes and two-way spike feet.

Due to the nature of electrical work, all ladders should be of a nonconductive type. Fiberglass ladders are the preferred choice for the trade. These ladders are expensive and heavy, but they are far safer than aluminum ladders. Check the weight rating for each ladder before you purchase it. There are many times when people buy ladders rated for a weight load of much less than what is needed. If you or your workers are heavy, the ladders you buy will have to support the weight. This is a big safety issue.

Having the right tool is a great start, but you must use the tool properly to ensure your personal safety and the safety of those around you. A great many accidents on jobs are related to the use of ladders.

Before using any ladder, check it for damage. Do a good visual inspection on your ladders before you use them. Set your ladders up properly. In the case of stepladders, make sure that the legs are set on a solid surface and that the side braces are fully engaged.

When setting up an extension ladder, make sure that the ladder you are using is the right size for the job you are doing. The ladder should be set up on level, solid surfaces. You may have to block one of the legs to accomplish a level setup. Tying the ladder off at its center point and at the top of the ladder is good safety advice. This will take a little extra time, but it's good insurance. The use of a safety belt may also be advisable.

KNOCKOUT PUNCHES

Knockout punches, excluding the left-hook type, come in manual and hydraulic designs. Manual punches are used in cramped spaces. They are used to punch new holes in panels, boxes, meter sockets, and so forth. A manual punch is operated with a wrench or a ratchet and socket. Sizes for these punches range from ½ inch to 5 inches.

Hydraulic Knockout Punches

Hydraulic knockout punches are much heavier and far more bulky than manual punches. But, they make punching holes in metal much easier when they can be used.

Ratchet Knockout Punches

Ratchet knockout punches are faster than the basic manual style of punch. The same punches used with basic manual punches and hydraulic punches can be used in conjunction with ratchet knockout punches.

One-Shot Punch

A one-shot punch is hydraulically operated. The big advantage to this tool is that it does not require you to drill starter holes for various sizes of punches. These punches are available in three sizes: ½ inch, ¾ inch, and 1 inch.

Stud Punch

A stud punch is used to make holes in metal framing studs. This device is a one-shot design that makes holes for ½ inch conduit and wire. The stud punch is operated by hand. Two handles of about 20 inches in length are used to operate the punch. The handles are opened and the punch is placed over the metal stud that is to be perforated. When the handles are squeezed together, a hole is created.

DRILL BITS

Drill bits come in various sizes and designs (Figures 11.1, 11.2). Electricians use numerous types of drill bits regularly. Some of the bits used are up to 72 inches in length. Bits are needed to drill wood and metal. There are also times when masonry bits are required for drilling brick, block, and concrete. An experienced electrician will have a wide assortment of drill bits available for any type of need encountered on a job.

HOLE SAWS

Hole saws are frequently used to create holes in metal, plastic, and wood. These devices range in size from about ⁹⁄₁₆ inch to 6 inches. Some electricians use these drill-operated hole saws to bore holes in everything from entrance cables to dryer vents. Other electricians prefer to use saw-type drill bits for this work. One drawback to hole saws is the need to clean them after each use. Getting the drilled material out of a hole saw can be time consuming and frustrating.

Wood Boring Bits

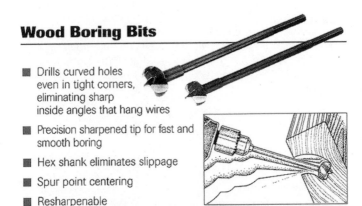

- Drills curved holes even in tight corners, eliminating sharp inside angles that hang wires

- Precision sharpened tip for fast and smooth boring

- Hex shank eliminates slippage

- Spur point centering

- Resharpenable

Description	RPM in Soft Wood	RPM in Hard Wood	Cat. No.
7/8 in. Wood Boring Bit	650-1000	600-850	**90-193**
1-1/8 in. Wood Boring Bit	650-1000	600-850	**90-196**

FIGURE 11.1 Wood boring bit designed for drilling curved holes that help eliminate sharp angles. *(Courtesy Ideal Industries)*

HACKSAWS

Hacksaws are available in many sizes and styles. This is a tool that is used regularly in the electrical trade. While the blades used in a hacksaw are the most important part of the tool, the hacksaw itself can help or hinder your work. Get a strong, durable hacksaw. Some models offer the ability to mount a blade in different locations on the hacksaw frame. This feature can come in handy.

Miniature hacksaws use regular hacksaw blades, but the saw is small and can be used in places where a standard hacksaw would never fit. No toolbox should be without a miniature hacksaw.

KEYHOLE SAWS

Keyhole saws pack a lot of cutting power in a small package. Whether cutting drywall or sawing through plaster lathe, a keyhole saw works well. These saws see a lot of use in the electrical trade and belong in every toolbox.

FIGURE 11.2 Flexible drill bits are available in many of styles for a variety of applications, as illustrated in this photo. *(Courtesy Ideal Industries)*

KNIVES

A good knife is an electrician's friend. Many types of knives are used in the trade. Knives are used to strip sheathing and insulation from wire. They are also used to scratch marks into metal objects, to locate drill points, and to ream conduit. A knife with a blade that locks into place is safer than one that does not have a locking feature.

ELECTRICIAN'S HAMMER

What makes an electrician's hammer different from a standard hammer? An electrician's hammer has a fiberglass handle and a longer snout that allows the hammer to reach into deeper recesses (Figure 11.3). The striking point on this type of hammer is small, only about one inch in diameter. These hammers work very well on staples and nails.

PLIERS

Pliers are needed daily in the electrical trade (Figure 11.4). A wide assortment of pliers should be in an electrician's toolbox. The uses of pliers in the trade include:

- *Cutting*
- *Holding*
- *Tightening*
- *Bending*
- *Crimping*
- *Locking*
- *Loosening*

FIGURE 11.3 Fiberglass handled electrician's hammer.

SCREWDRIVERS

Screwdrivers, like pliers, are used daily by electricians. An assortment of screwdrivers is needed (Figures 11.5, 11.6). The type that has a head that can be changed from a flat bit to a Phillips head is very convenient. Short screwdrivers are sometimes needed. Long screwdrivers can also come in handy. Bits on screwdrivers are needed in many different styles and sizes. Ratchet-style screwdrivers can be real timesavers. Electric screwdrivers work wonders, as well. Screwdrivers that are designed to hold screws to the bit reduce the risk of losing screws and can make a job go faster.

MULTI-TAP TOOLS

Multi-tap tools offer three taps on a screwdriver handle (Figure 11.7). The tool is used to re-thread stripped screw holes, to clean plaster from screw holes in electrical boxes, and to tap metal to attach pipe straps.

FIGURE 11.4   Insulated electrician's pliers.

FIGURE 11.5   Insulated electrician's screwdrivers.

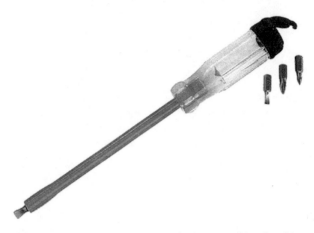

FIGURE 11.6   Five-in-one non-conductive screwdriver/nutdriver.

FIGURE 11.7   Triple-tap tool taps three hole sizes, and cleans and chases burred threads.

NUT DRIVERS

Nut drivers can make some jobs go together quickly. An assortment of these tools will make working with nuts easier and faster. Nut drivers can get into spaces where a normal wrench could not be used.

ALLEN WRENCHES

Allen wrenches are used for many purposes. They are often used when working with circuit breakers, lugs, and wire-splicing clamps. An electrician's toolbox should contain a variety of Allen wrenches.

TAPE MEASURES

Tape measures are an essential tool. They come in various sizes, but most pros use power-return tape measures in lengths of 16 feet, 20 feet, 25 feet, or 30 feet. You should have a few tape measures on hand, because if one becomes broken or lost, your workday could be ruined until you can make a run to the supply house.

TORPEDO LEVELS

When buying a torpedo level, get the type that has a magnetic strip on the bottom edge of the level. These levels are small, about 9 inches long, but they are essential to the trade. Torpedo levels are used in the installation of boxes and conduit to keep the items plumb and level.

PORTABLE BAND SAWS

Portable band saws perform many duties on electrical jobs. These saws are basically electrical hacksaws that can be used to cut wood, plastic, and metal. Cutting iron, large conduits, stainless steel, large electrical cables, and even some raceways are no problem with a portable band saw. These saws can work in close conditions and are used frequently on jobsites.

RECIPROCATING SAWS

Electric reciprocating saws see a lot of action in the hands of an electrician. With interchangeable and disposal blades, these saws can cut wood, metal, and plastic. Always keep a wide assortment of blades on hand. The blades come in many different lengths and styles.

CONDUIT BENDERS

Conduit benders range in size from ½ inch to 1¼ inch (Figure 11.8). These tools are used to bend conduit and they can produce bends of 90 degrees.

Rigid hand benders come in several sizes. They are segment benders that are used to bend rigid, not EMT, conduit. These benders can require two people to operate, since they will bend rigid conduit that has a large diameter. A 3 inch conduit can be bent with this type of tool.

One-shot, or hydraulic, benders can bend conduit that has a diameter of up to 6 inches. Hydraulic benders can be used in conjunction with hand pumps or electric pumps. Some of the bends possible with these units include kicks, offsets, 45 degree bends, and 90 degree bends.

METERS

An electrician would be lost without electrical meters. There are many types of meters used by electricians (Figures 11.9 through 11.11). Some of the most common types of meters include the following:

- *Voltage meters*
- *Volt-Ohm testers*
- *Volt-Ohm-Amp meters*
- *Capacitor testers*

A volt meter is used to check voltage from one hot leg to neutral or ground or across both hot legs. Handheld meters of this type measure voltage from 120 volts of alternating current (AC) to 550 volts. A handheld volt tester can read direct current (DC) in the range of 120 volts to 600 volts.

FIGURE 11.8 Detail of conduit bender head.

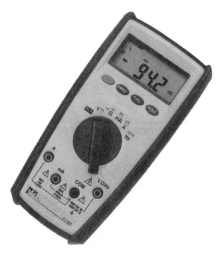

FIGURE 11.9 Commercial grade multi-meter.

FIGURE 11.10 A clamp meter rated to 1,000 amps.

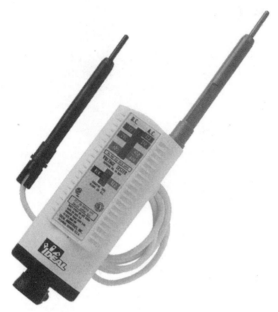

FIGURE 11.11   Votage tester.

Volt-Ohm-Amp Meter

A volt-ohm-amp meter, also know as an amprobe, is a multi-function meter that can be used as an AC volt meter and a continuity tester. The jaws of an amprobe are opened by pressing a lever on the side of the device. Once the jaws are clamped around one wire, the amp load can be tested. This test is done with one wire at a time. Amprobes can also be used on poly-phase loads to balance the load centers of a job.

Volt-Ohm Tester

A volt-ohm tester is battery operated and is used to measure voltage, resistance, and current. These devices can be used with both AC and DC current. A tester of this type can be used to ring out and number wires. Measuring resistance of wires to find shorts and grounds or a break in wires is another function of a volt-ohm tester. Additionally, the tester can measure amps up to 5 amps.

Capacitor Testers

Capacitor testers are battery operated, portable testers that can do a quick test on a capacitor with an audible signal. The tester tests the capacitor to determine if it is open, shorted, or still in good condition.

CORDLESS TOOLS

Cordless tools are extremely popular in modern construction. These battery-powered tools are very convenient. Some of the cordless tools that are available include:

- *Flashlights*
- *Drills*
- *Screwdrivers*
- *Reciprocating saws*
- *Circular saws*

PIPE THREADERS

Pipe threaders are available in manual models and electric models. Ratchet threaders are operated by hand and don't require electrical power. These threaders can thread pipe with a diameter of up to 2 inches. An electric handheld threader is available and it uses the same dies as those used by a ratchet threader. These threaders take a lot of the labor out of threading pipe when electricity is available.

Larger, electric pipe threaders can thread pipe with a diameter of up to 6 inches. A pipe vise, like a tripod vise, is needed when working with a large electric threader. Any electric threader can pose a safety risk. Never wear jewelry or loose-fitting clothing when operating an electric threader.

In addition to a pipe vise, a pipe cutter and a pipe reamer are needed when working with heavy metal pipe. An oiler is another accessory that is needed when threading pipe. Oiling the pipe as it is being threaded protects the dies from damage. The edges of cut pipe have to be reamed to a smooth condition to avoid damaging wires that will be pulled through the pipe. Some large threaders come equipped with all the needed accessories as a part of the tool.

FISH TAPES

Fish tapes come in many lengths. The most common fish tapes range in length from 50 to 200 feet. Most fish tapes are either ⅛ inch or ¼ inch wide. Fish tapes are used to pull wires through conduit and walls that are closed. Good fish tapes are housed in a spool with a crank (Figure 11.12). The spool prevents the fish tape from becoming tangled and bent. Fish-tape leaders can be attached to the end of a fish tape. The leader helps to maneuver a fish tape around bends and makes pushing the fish tape through a conduit easier.

WIRE LUBE

Wire lube can be used when pulling wire through a conduit. This is a special lubricant designed for use with electrical wires. Don't sub-

FIGURE 11.12 Fish tapes housed in spools.

stitute soap, oil, or grease as a lubricant. These materials can degrade the sheathing and installation of wires. Use only approved lubricants for pulling wires (Figure 11.13). You can buy wire lube in quart cans or five gallon pails. The lube is also available in quart squeeze bottles and aerosol cans. The wax-like lube is very slippery and distributes easily over wires. If you will be pulling long lengths of wire, your work will be much easier if you apply wire lube prior to pulling the wire.

ELECTRICAL CONNECTORS AND ACCESSORIES

When you get into the subject of electrical connectors and accessories, you are into a vast array of devices that make electrical installations faster, easier, and safer. A lengthy volume could be written on the subject of connectors and accessories. Manufacturers are happy to provide licensed electricians with catalogs of their products. It is

FIGURE 11.13 Approved wire pulling lubricants, as shown in this photo, are safe to use with all cable types, and are environmentally safe.

a good idea to request catalogs from many manufacturers. Once you have them, build a reference library. You will find the information in the catalogs enlightening. Tools, connectors, and accessories for the electrical trade are numerous. The best way to keep up with these items and new developments is to stay in touch with your suppliers and the manufacturers of various products.

12

TROUBLESHOOTING
AND REPAIRS

G ood troubleshooting skills are very important for electricians
who do service and repair work. Electricians who specialize
in new construction may not need to be terribly concerned
with troubleshooting skills. However, many electricians specialize in
service and repair work. There are also many electricians who are
generalists and who need strong troubleshooting skills to make a
good living.

Electricians who do service work and who don't understand the
importance of troubleshooting skills are likely to make a bad impression on their customers. This is a good way to lose business. Bumbling
around with a hit-or-miss procedure won't cut it in the professional
world. No, you have to be systematic in your search for the cause of
a problem.

Troubleshooting skills develop with experience, but you can jump-
start your skills by learning from others. Working closely with a seasoned pro who is willing to share techniques with you is the best way
to pick up superior troubleshooting skills. The earn-while-you-learn
plan works, but it can take a considerable amount of time to hone
your skills. Reading is a very effective way to perfect troubleshooting
skills, and that is why you are reading this chapter.

PLUGS AND CORDS

Plugs and cords are small and may seem insignificant, but plugs and cords are a major cause of electrical shocks and fire hazards. These devices can be inspected for loose wires, damaged connections, compromised insulation, and brittle prongs. These are simple inspections to perform. If a plug is defective or a connection is risky, you can simply replace the existing plug at minimal expense. This may not sound like much of a job for a licensed electrician, but even simple repairs are important.

Replacing Round Plugs

Replacing round plugs is simple. There are only five steps involved in the job. The first step is to cut the old plug off the cord. Remove the deadfront cover from the prong end of the new plug. Slide the cord through the plug and strip off about 3 inches of the sheathing. Then strip about ½ inch of the wire insulation.

This next step is a bit of a trick of the trade. Tie a knot in the two wires where the insulation remains. This will keep the wire from being pulled out of the plug if the cord is tugged on. Bend clockwise hooks in the ends of the bare wire. Secure the black wire under the brass screw and then secure the white screw under the remaining screw. Replace the deadfront cover and make sure that all wires and strands are completely concealed in the plug and under the deadfront cover. That's all there is to it.

240 VOLT PLUG

Replacing a 240 volt plug is no big deal. This type of plug has a steel clamp that grips the cord. Cut off the old plug. Loosen the retainer clamp on the new plug, remove the deadfront cover, and slide the wire through the plug. Strip out about ½ inch of wire insulation. Make sure the wire strands are twisted tightly. Use a clockwise hook on each bare wire. Attach the red and black wires under brass screws. The green wire should be installed under the green screw. Tuck the wires in tight and replace the deadfront cover. You're done.

QUICK-CONNECT PLUGS

If you want to replace a plug on a flat cord, use a quick-connect plug. This job is almost too simple. Cut the old plug off. Raise the lever on the quick-connect plug. Slide the zip cord into the plug and press down firmly on the lever that is on the top of the plug. When you do this, the zip cord is pierced and the connection is completed.

INCANDESCENT FIXTURES

There are countless types of incandescent fixtures, but, fortunately, most of them are put together in a similar way. This makes troubleshooting them much easier than if they were all different. When a fixture shorts out and blows a circuit breaker or fuse, the problem is probably in the fixture. If the light won't come on at all, it's probably either a bad light bulb or a bad switch.

If you suspect a problem with a light fixture, check the socket. It may be cracked. Make sure the switch to the fixture is turned off before you go poking around. If the socket looks okay, check the wires. Are they burned? When everything looks fine, you will have to remove the light bulb and check the contact at the socket base. There could be corrosion on the contact. If there is, scrape it off with a flat screwdriver or some steel wool. Try prying up the contact a bit, to ensure good contact. Once this is done, put the unit back into operation and test it. If the problem persists, you will have to move onto checking the wiring. If light corrosion is present in the lamp socket, apply a thin coating of antioxidant compound to the inside of the screw shell. You can also apply a light coating to the threads to the base of the bulb. This will allow better contact and full seating of the lamp base. This is a good preventive maintenance measure for exterior fixtures.

When you have to check the wiring, cut off the power to the fixture. Loosen or remove the mounting screws and lower the fixture

Trade Tip: Never leave a service panel exposed without its cover in place. It takes only moments for inexperienced people to receive serious, and possibly fatal, shocks from exposed bus bars.

from the box. Are there any loose connections? Is the wiring insulation in tact? Has the fixture been overheating? If the wire insulation is cracking, it can be a sign of overheating. Peeling drywall paper near the fixture might be an indicator that the fixture is overheating. If the fixture is overheating, it is either defective or it has been housing a bulb with a wattage rating that is too high for the fixture.

Depending upon what you find, you might make a minor correction, such as fixing a loose connection. The repair could be as simple as putting in a new light bulb that has a lower wattage rating. In the worst case, replace the fixture.

FLUORESCENT FIXTURES

Fluorescent fixtures are considerably different from incandescent fixtures. The ballast in a fluorescent fixture is a key element to tune in to when you are troubleshooting the fixture. Incandescent light bulbs just burn out. Fluorescent light tubes rarely go out in the blink of an eye. They tend to die out slowly. If you have a fluorescent light that will not come on, inspect the light tubes. A tube that has been in use will have some graying on the tube, near the contact ends. A worn-out tube will have black discoloration on the tube. If the tube looks clean, wiggle the tube to see if the contact points are making good contact. This will often solve the problem. If not, take the tube out and put it back in before you go into deeper troubleshooting. The best way to determine whether or not the problem is with the tubes is to try them in a fixture that is known to be working. Likewise, take the tubes from the working fixture and try them in the problem fixture.

Some fluorescent fixtures are equipped with replaceable starters. These are generally older fixtures. It is common for these types of fixtures to flicker when they light up. If the flickering continues for an extended time, the starter might not be seated properly. Push in on

Did You Know: that a multitester fills the role of both a voltage tester and a continuity tester? It does. This tool can also be used to test low-voltage wiring. Most pros prefer digital testers, rather than those with needles.

Trade Tip: Keep more than one tester on your truck. You never know when a tester will fail to function. The battery might be dead. A fuse might blow. You could lose a tester. Make sure you have a backup tester so that you are not in the field without the ability to test systems.

the starter and turn it clockwise. This could be all that you have to do. If the tube lights up at each end and doesn't light up in the middle, it is an indicator that the starter is bad. You can replace the starter by pushing in on it and turning it counterclockwise.

If a fluorescent fixture hums or emits a material that looks dark and sticky, the ballast probably needs to be replaced. In many cases, it is easier and not much more expensive to replace the entire fixture. If you opt to replace the ballast, turn off the power to the unit and remove the light tubes.

You can remove a ballast by releasing the wires at the sockets. This is done by pushing a screwdriver into the release opening. Unscrew the ballast and disconnect the wires to the power source. Install a new ballast and try putting the unit back into operation. If it doesn't work, trace backwards into the fixture box and confirm that you have good power coming to the fixture. Many seasoned electricians would do this prior to replacing the ballast, but some will replace the ballast first, since the symptoms usually indicate a bad ballast.

TROUBLESHOOTING SWITCHES

Switches wear out. If a switch isn't working, you can check for bad connections, but you might as well just swap out the switch. This is not an expensive process and it ensures a new switch that should give years of use and help you avoid a callback.

Neon Tester

A neon tester can be used to test a switch. Turn off the circuit breaker or remove the fuse that controls the power flow to the switch. Yeah, I know, who bothers? Well, that's up to you, but if you want to be safe, kill the power. Since you are a pro, you can probably avoid being lit up when working with the power on, but I have to tell you the safe way to do it.

First remove the coverplate and the retaining screws and pull the switch assembly out of the box. Once the switch is out of the box and the contacts and bare wires are in a safe location, turn the power to the switch on. The switch should be in the off position. Touch the probes of the neon tester to the switch screws. If the tester glows, you have power to the switch. Turn the switch on and touch the probes to the screws again. If the tester glows with the switch in the on position, the switch is faulty and must be replaced.

To replace the switch, cut the power off to the switch. Loosen the connection screws and remove the defective switch. Attach the wires to the connectors on the new switch and install the switch in the box. Replace the coverplate and turn on the power. Test the switch and everything should work.

Continuity Tester

A continuity tester can be used to test switches. Turn the power off to the switch and remove it from the box. Disconnect the wires on the switch. Attach the clamp of the continuity tester to one screw and place the probe on the other screw. Your tester should glow when the switch is in the on position and it should not glow when the switch is in the off position. This will tell you if the switch is faulty.

fastfacts

Continuity testers should never be used on hot wires. All power should be turned off to wires before checking for continuity. A battery in the continuity tester provides all the power that is needed to conduct a successful test. If you connect your continuity tester to a live circuit you will blow your tester. The smartest thing to do before testing for continuity is to test for voltage. If no voltage is present on the circuit, you are good to go; but if voltage is present, you need to de-energize the circuit that is to be tested.

If you are going to use a continuity tester to test a three-way switch, shut off the power to the switch. Attach a clip to the common terminal. If you are confused as to which is the common terminal, look at the device. It should be labeled. Touch the probe to one of the other screw terminals and activate the switch. If the tester lights up, the switch is okay. Now put the probe on the other terminal and repeat the test. If the tester doesn't light up when on either of the terminals independently, the switch is bad.

You can also use a continuity tester to test a combination switch-receptacle. To do this, attach the clip to one of the top switch terminals and touch the probe to the top terminal on the other side. Remember that there should be no wires attached to the switch during the test. If the light on the tester comes on when the switch is in the on position, the switch is fine.

TROUBLESHOOTING RECEPTACLES

Troubleshooting receptacles is easy. All you need is a neon tester and possibly a screwdriver. With the power to the outlet on, insert a probe into each opening of one of the outlet connections. Keep your fingers off of the bare probes and handle only the insulated sections. Remember, the power is on. If the tester lights up, the outlet is okay. When the tester doesn't light up, you need to remove the coverplate from the outlet. Place the probes on the screw terminals. If the tester lights up, you have power to the outlet and a defective receptacle.

In the event that you encounter a defective outlet, cut the power off to the outlet. Loosen the terminal screws and disconnect the wires from the outlet. Install the wiring on a new outlet and secure the outlet in the box. Replace the coverplate and turn the power on. Test the outlet with your tester. It should work.

 Trade Tip: Have you ever had wires slip between your fingers and fall into a wall cavity? If you have, you know how frustrating this situation can be. Avoid this by putting a clip on the wire insulation to prevent the wire from receding into a hole.

fastfacts

Working in older homes where there is plaster and lathe can be a real pain. Installing new boxes in plaster walls can be frustrating. One way that works well is to cut a hole in the wall and lathe that is a little larger than the box. Once the box is adjusted for installation depth, you should drill small pilot holes in the lathe. Without the pilot holes, the lathe may break when screws are installed. Allow the mounting screws to enter the pilot holes and you should avoid this problem. A neat method of mounting boxes in lathe and plaster is to place them next to a stud and secure them directly to the wall stud.

Polarization

You may have an occasion to test an outlet for polarization. With the power to the outlet on, put one of the tester probes into the hot slot of the outlet. The hot slot will be the shorter of the two. The neutral slot is the longer of the two. The other probe should be inserted into the grounding plug hole. If the tester lights up, the outlet is polarized and grounded. If the tester doesn't glow, leave one probe in the grounding hole and insert the other probe in the long outlet slot. A tester that glows under these conditions indicates that the hot and neutral wires are reversed.

If the tester doesn't glow during either test, the outlet is not grounded. Another simple test is to check for continuity between the ground slot and the neutral slot. This procedure is normally done with the power off because connecting a continuity tester to a live circuit can damage your tester.

 Don't Do This! Circuits should carry loads that do not exceed 80 percent of the circuit's suggested maximum rating. For example, a 20 amp circuit should not be loaded with more than 16 amps. Don't load a circuit to its maximum capacity.

TROUBLESHOOTING CIRCUIT BREAKERS

Circuit breakers can go bad. They can also be tripped for problems that might turn serious. The basic function of a circuit breaker is to protect against shorts and overloaded circuits. They do this by sensing heat. A bimetal strip provides tension to keep a breaker in a functioning position. When the bimetal strip detects a short or an overheated circuit, the strip heats up and bends. At this point, the breaker is tripped to cut power off to the suspect circuit. The circuit will remain off until the breaker is reset.

If a circuit is dead, check the panel box to see if any breakers are tripped. When a tripped breaker is found, reset it. If it trips again, you have some work in front of you. A breaker that trips repeatedly when a circuit is not overloaded indicates a short. You can determine if a circuit is overloaded by doing a load rating of the devices connected to the circuit.

A defective plug, cord, or light socket could create a short that would trip a breaker. Check for these types of problems. Inspect connections in switch boxes and outlet boxes. You may find conditions in the box that would cause a short. Look at the wiring in the boxes. If it is frayed or shows damaged insulation, you may have found your problem. If this is the case, your choices are to repair the damage by cutting the wire that is damaged and replacing it with new wiring. Use a heat-shrink tube to reinsulated the wire, or you might use electrical tape. The best way is to cut out the bad wire and replace it with a jumper.

Don't forget to check the wattage ratings of light fixtures and light bulbs. If a bulb is installed that has a higher wattage rating than what a fixture is rated for, overheating can occur, and this can damage insulation on electrical wires. This, too, could cause a breaker to trip.

fastfacts

Don't be tempted to upgrade a fuse if you are dealing with a circuit that is being overloaded. Installing a fuse with a higher rating than what is intended for a circuit can result in a fire. Never install a fuse that is rated higher than it should be for the circuit being served. The fuse size protecting the circuit is limited to the ampacity rating of the wire serving the circuit.

fastfacts

Older wiring that has developed nicks and scratches in the bare wiring pose a risk of overheating. If you discover this type of situation, clip the bare wire back to the insulation, strip some of the insulation and make a new connection with undamaged wire.

By the way, the same basic troubleshooting steps discussed here will work with fuses. It is worth noting that when working with fuses, a fuse that has blown due to a short will usually have an exploded strip in the inspection window. The fuse's inspection window will normally be blackened due to a short. If the cause of the problem is overheating, the inspection window should be clear and the strip should be melted.

If you want to test a cartridge fuse with a continuity tester, pull the fuse with a fuse puller. Attach the continuity tester on one end of the fuse. Touch the probe of the tester to the other end of the fuse. If the tester lights up, the fuse is okay.

DOORBELLS

Doorbells occasionally fail. Most doorbells incorporate a transformer that converts incoming 120 volt power to somewhere between 6 and 30 volts. A small bell wire, often something in the range of 18 gauge wire, connects the bell, chime, or buzzer to the back of a circuit at the doorbell. You can troubleshoot a doorbell without cutting off the electricity to the circuit, unless you are working with the transformer. The troubleshooting process is a matter of elimination in organized steps.

The first thing to check is the button that operates the doorbell. Remove it and place a clamp on the wires to prevent them from slip-

Did You Know: that the ends of cartridge fuses get hot? They. do, so don't touch them when they are first pulled out of a service panel.

ping into the siding or trim of a building if they should become disconnected from the button. Next, touch the two wires together momentarily. If the bell rings, you know the problem is with the button. If there is a corrosive buildup on the wires and connections, clean it off and test the doorbell. If the bell does not function, make sure that the electrical connections are tight and test the bell. Make sure that the wiring insulation is not damaged.

Buttons are often the root of doorbell problems. You can test a button by jumping the terminals with a short piece of bell wire. If the bell rings when you apply the jumper to the terminals, you have determined that the button is defective. If you prefer to use a continuity tester, disconnect the wires from the button. Attach the tester clamp to one terminal and place a probe on the other terminal. Press the button. If the tester glows, the button is fine.

If the button checks out, you will have to check the transformer. You should turn off the power to the transformer before working with it. Inspect all of the wiring for corrosion, loose wires, and damaged insulation. Disconnect the wires from the transformer. You will need a voltmeter or multitester for this test. Put a probe from the tester on each terminal of the transformer. Set the meter dial to ACV 50. Turn the power on to the transformer. If the meter doesn't show a reading of voltage, the transformer is bad. If you have to replace the transformer, make sure that the power to the device is off. The replacement process is simple. Just remove the old transformer and install the new one on the existing wires in the box.

When you have followed all of these instructions and the doorbell still doesn't work, you have to check the bell. Remove the bell

fastfacts

What does it mean when you see rusted areas around the bases of fuses in a service panel? It is generally evidence of water seeping into the box. Obviously, this is not a good situation. A full inspection of the service head and main service panel is needed under these conditions. The source of water entry has to be resealed to eliminate moisture in the service box. Excessive rust and corrosion generally means that you need to replace these components of the service entrance.

fastfacts

As a reminder, the main breaker or fuse in a service panel can be cut off or removed without de-energizing the main power cables coming into the panel box. Don't allow a careless mistake to ruin your future.

cover and do a visual inspection for bad connections. If you don't see any, it's time to put your meter back to work.

Most bells are set up for two buttons. Put the probes from your meter on the front and transformer terminals. If the meter shows a reading, it means that power is going to the bell. This would indicate that the bell is defective. Check the terminals for the second button to make sure that it reads okay. By now, you should have found the problem. If not, you may be faced with a broken wire somewhere within the system. To check the wiring, disconnect the wires going to the bell and test the current flow of the wiring with your meter. If there is a break, you will have to replace the wiring.

THERMOSTATS

Thermostats generally suffer from one of three potential problems. Faulty wiring, corrosion, and worn-out transformers are the most common reasons for thermostat failure. Dust in a thermostat can also cause the device to malfunction. This is a good place to start your troubleshooting.

When you are dealing with a thermostat that is not working properly, remove the cover plate and use a can of compressed air to

Trade Tip: How would you repair insulation on a wire that has cracked, broken, or been cut? The damaged insulation can be repaired with heat-shrink tubing. The tubing covers the damage area and is then heated with a heat gun to shrink the tubing and form new insulation.

Did You Know: that aluminum wire that is all-aluminum can pose hazards when attached to a switch or outlet via a brass or copper screw terminal? It can. The aluminum may expand and contract at a rate different from that of copper or brass. This could result in a loose connection and possibly a fire.

blow out any dust build up. If dust doesn't seem to be the problem, check to see if the base of the thermostat is securely attached to the wall. If it is not, the thermostat may be tilted, and this could cause a problem. Check the base to see if any wires appear to be loose or corroded. Make sure that all terminal screws are tight.

If you have done the simple troubleshooting steps and have had no success, you can use a jumper between the terminals to test the base. Strip insulation from the short jumper wire at each end. Then put one end of bare wire on the R terminal. Place the other bare end on the W terminal. The unit controlled by the thermostat should come on. If it does, move the jumper to the Y terminal and the G terminal. This should turn on the fan of the unit controlled by the thermostat. If these elements do turn on, the thermostat is bad and needs to be replaced.

You can also test the transformer with your meter, if needed. Put one meter probe on each of the low-voltage terminals of the transformer. Set the dial to ACV 50. The meter should show power if the transformer is okay. Otherwise, you need to replace the transformer, assuming that power is coming to it.

fastfacts

What do you do when you are faced with short wires in an existing box? This is not a big deal. Cut a couple of sections of wire that is of the same gauge as the existing wire and make a jumper connection to give you longer wire to work with. These jumpers can be connected to existing wiring with wire nuts or compression crimp sleeves.

Before you replace the transformer, make sure that it has power coming to it. Check the power feed with your meter. Do this by opening the transformer box and putting a probe from the meter on the hot wires and the other to the box, if the box is grounded. If the box is not grounded, put a probe on the neutral wire. Assuming that you have power coming to the transformer and the transformer is not showing a good reading, replace it.

Troubleshooting is a progressive skill. The more troubleshooting you do, the better you will get at it. There are, of course, other topics that could be covered when it comes to troubleshooting electrical devices. Unfortunately, space is limited here and there is not enough room to cover every possible situation. What you have been given here will cover the most common needs. The logistical steps you have learned here can be applied to all forms of troubleshooting. Take the time to learn how to find the cause of problems without wasting your time and your customer's money.

fastfacts

How many times have you seen wires spliced together and hanging out in the open? Probably more often than you would have imagined. Do not do this. Use a junction box to house the splice. Exposed wires, even when they have wire nuts and tape on them, are an unnecessary risk. All splices must be contained within an approved and listed covered box.

GETTING WORK

Building business clientele is one of your most important jobs as a business owner. Learning how to win bids is one of the most effective ways to build your business clientele. How many times have you bid a job and never heard back from the potential customer? Many contractors never figure out how to win bids successfully. This chapter is going to show you how to win bids and build up your business.

WORD-OF-MOUTH REFERRALS

Word-of-mouth referrals are the best way to get new business. Of course, you will need some business before you can benefit from word-of-mouth referrals, but every time you get a job, you need to work that customer for referrals that will lead to more jobs.

Getting referrals from existing customers is not only the most effective way to generate new business, it's the least expensive. Advertising is expensive. For every job you get from advertising, you are losing a percentage of your profit to the cost of advertising. If you can turn up new work by talking with existing customers, you eliminate the cost of advertising.

If you do good work and take care of your customers, referrals will be easy to get, but you may have to ask for them. People will sometimes give your name and number to friends, and they occasionally write nice letters. However, to make the most of word-of-mouth referrals, you have to learn to ask for what you want. Let's see what it takes to get the most mileage out of your existing customers.

GROUND WORK

Laying the groundwork is an important step in getting a strong portfolio of customer referrals. If you don't make your customers happy, they are going to talk to their friends, but they won't be saying what you want to hear. People are fast to reveal their bad experiences, but they are not so quick to spread good words. To deliver the message to potential customers, you have to work hard at pleasing your present customers.

To lay the groundwork, you must start with the first contact you have with customers and maintain it right through to the end. Many contractors start off on the right foot, only to stumble before the job is done. I have seen many jobs go sour in the final days of completion. One of the largest mistakes I have seen contractors make is not responding promptly to warranty calls. If you are in business for the long haul, you don't want to alienate customers, even after the job is done. Old customers often become repeat customers. If you don't respond to callbacks, you won't be called when there is new, paying work to be done.

DURING THE JOB

During the course of a job you must cater to the customer. Most contractors don't have a problem with this aspect of customer satisfaction, but there is more to making customers happy than doing good work. You have to fulfill your promises, be punctual, be respectful, and be professional.

AT THE END OF THE JOB

At the end of the job you have to ask for referrals. Don't expect customers to run up to you and hand you a letter of reference. Often, asking for a letter of reference isn't enough. It helps if you provide a form for the consumer to fill out. People never seem to know what to say in a reference letter. They are much more comfortable filling out forms. If you design a simple form, almost all satisfied customers will complete it and sign it. I'm sure you have seen these quality-control forms in restaurants and with mail-order shipments. You can structure the form in any fashion you like.

Once you have designed and printed your forms, use them. When you are completing a job, ask your customer to fill out and sign your reference form. Do it on the spot. Once you are out of the house, getting the form completed and signed will be more difficult.

As you begin building a good collection of reference forms, don't hesitate to show them to prospective customers. Use an attractive three-ring binder and clear protective pages to store and display your hard-earned references. When you get enough reference letters, you will have strong ammunition for closing future deals.

CUSTOMER SATISFACTION

Customer satisfaction is a key to success. Business builds upon itself when customers are satisfied. Oh sure, there are some people that you may never be able to satisfy. These hard-to-please people seem to exist for every business owner. If you haven't run into a habitual complainer, you will, if you're in business long enough. Leaving this small segment of the population out of the picture, let's concentrate on how to please the bulk of your customers.

Customers like to feel comfortable with their contractors. To make customers comfortable, you have to deal with them on their level. You will have to learn how to play the give-and-take game. Communication skills are essential to a good relationship. If you and the customers can't communicate, you won't get far in your business dealings. Occasionally, you will have to baby-sit them. You might not like having to soothe customers, but there will be times when you have to smooth the feathers of ruffled clients.

NEW CUSTOMER BASES

You can start reaching out for a new customer base through bid sheets. Bid sheets are open to all reputable contractors. When a job is placed on a bid sheet, somebody is going to get the job. Also, a job put out to formal bids is almost always a known quantity. Unlike common residential estimates, where the potential customers change their minds, formal bid sheets are rarely changed. This type of work is very competitive and the percentage of profit is usually low, but bid work can pay the bills.

What Is A Bid Sheet?

What is a bid sheet? A bid sheet is a formal request for price quotes. There is a difference between a bid sheet and a bid package. The bid sheet will give a brief description of the work available. A bid package gives complete details of what will be expected from bidders. Most contractors start with a bid sheet and, if they find a job of interest, order a bid package. Bid sheets are usually provided free of charge. Bid packages often require either a deposit or a non-refundable fee.

Where Do You Get Bid Sheets?

Where do you get bid sheets? Bid sheets can be obtained by responding to public notices in newspapers and by subscribing to services that provide bid information. If you watch the classified section of major newspapers, you will see advertisements for jobs going out for bids. You can receive bid packages by responding to these advertisements. Normally, you will get a set of plans, specifications, bid documents, bid instructions, and other needed information. These bid packages can be simple or complicated.

Bidder Agencies

Bidder agencies are businesses that provide listings of bid opportunities. These listings are normally published in newsletter form. The bid reports are generally delivered to contractors on a weekly basis. Each bid report may contain five jobs or fifty jobs. These publications are an excellent way to get leads on all types of jobs.

What Types Of Jobs Are On Bid Sheets?

What types of jobs are on bid sheets? All types of jobs can appear, ranging from small residential jobs to large commercial jobs. The majority of the jobs are commercial. The prices of the jobs range from a few thousand dollars to millions of dollars.

Government Bid Sheets

Government bid sheets are another opportunity for finding an abundance of work. Like other bid sheets, government bid sheets give a synopsis of the job description and provide information for obtaining more details. Government jobs can range from replacing a dozen light fixtures to building a commissary. Building new military base housing units could result in many months of work for a homebuilder. Electricians can cash in on repairs, remodeling, and new construction.

Government jobs are a safe bet for collecting your money. The money may be slow in coming, but it will arrive. The paperwork involved with government jobs can be excessive. If you are not willing to deal with mountains of paperwork, stay away from government bids.

ARE YOU BONDABLE FOR LARGE JOBS?

Are you bondable for large jobs? Many of the jobs found on bid sheets require contractors to be bonded. Bonds are obtained from bonding companies and insurance companies, but not all contractors are bondable. Before you try bidding jobs that require bonding, check to see if you are bondable. The requirements for being bonded vary. Check in your local phone book for an agency that does bonding, and call to inquire about the requirements.

PERFORMANCE AND SECURITY BID BONDS

Performance and security bid bonds are a necessity with many major jobs. If you order a bid sheet or package, you will almost certainly see that a bond is required. Some listings on bid sheets may not require a bond. It is common for bid requirements to be tied to the anticipated cost of the job. The bigger the job, the more likely it is that a bond will be required.

Why Are Bonds Required?

Why are bonds required? Bonds are required to ensure the success of the job. The people offering the work want to be sure that the job will be done right and that it will be completed. When the person or firm issuing the work requires a bond, they know there is a degree of safety. It is very difficult for some new businesses to obtain a bond. If the new company doesn't have strong assets or a good track record, getting a bond is tough. Unfortunately, a bond can be a hurdle you can't get over until you are so successful that you don't care whether you have it or not.

There are three basic types of bonds to be considered: bid bonds, performance bonds, and payment bonds. Each type of bond serves a different purpose. A bid bond is put up to assure the person receiving bids that the bidding contractor will honor the bid if a contract is offered. Performance bonds prevent contractors from abandoning a job and leaving the customer in dire financial straits. If a contractor reneges on completing the job, the customer may hold the performance bond for financial damages.

Payment bonds are used to guarantee payment to all subcontractors and suppliers used by a contractor. These bonds eliminate the risk of mechanic and materialman liens being filed against the property where work is being done.

When you put up a bond, the value of the bond is at risk. If you default on your contract, you lose your bond to the person that contracted you for the job. Since many people use the equity in their home for collateral to get a bond, they could lose their house. Bonds are serious business. If you can get a bond, you have an advantage in the business world. Talk with local companies that issue bonds to see if you can qualify for bonding.

THE RISK OF BIG JOBS

Are big jobs surrounded by big risks? You bet they are. There are risks in all jobs, but big jobs do carry big risks. Should you shy away from big jobs? Maybe, but if you go into the deal with the right knowledge and paperwork, you should survive and possibly prosper.

Cash-Flow Problems

Cash-flow problems are frequently present with contractors doing big jobs. Unlike small residential jobs, big jobs don't generally allow for contractors to receive cash deposits. If you tackle these jobs, you will have to work with your own money and credit. For a new business, the money needed to draw a disbursement in big jobs can be the undoing of the company. It's wonderful to think of signing a million dollar job in your first year of business, but that job could put your business into the bankruptcy court.

Before you dive into the deep water, make sure you can get to the other side. Some lenders will allow you to use your contract as security for a loan, but don't bet your business on it. If you want to take on a big job, get your finances in order first.

Slow Pay

Slow pay can be another problem with big jobs. Large jobs are notorious for slow pay. It's not that you won't get paid, but you may not get paid in time to keep your business going. New businesses are especially vulnerable to failure by slow pay. When you move into the big leagues, be prepared to hold your financial breath for awhile. The check you thought would come last month might not show up for another 90 days.

No Pay

Slow pay is bad, but no pay is worse. Just as new contractors can get in trouble with large jobs, developers and general contractors can get into financial difficulty with big jobs. Most of the people spearheading big jobs don't intend to stick their subcontractors, but sometimes they do. When the top dogs get in over their heads, they can't pay their bills. If the subcontractors don't get paid, suppliers don't get paid. The ripple effect continues. Anyone involved with the project is going to lose. Some will lose more than others. Generally, when these big jobs go bad, the banks or lenders financing the job will foreclose on the property. These lenders normally hold a first mortgage on the property.

If you're working as a subcontractor for a large outfit, filing a mechanics lien is the best course of action when your customer refuses to pay you. If a contractor hasn't been paid for labor or materials, a mechanics lien can usually be levied against the property where the labor or materials were invested. If you have to file a lien, make sure you do it right. There are rules you must follow in filing and perfecting a lien. You can file your own liens, but I recommend working with an attorney on all legal matters.

Even after you file and perfect your lien, you may not get your money. If you get any money, it will likely be a settlement in a reduced amount. You can never quite get the taste of sour jobs out of your mouth.

Completion Dates

Completion dates can also wreak havoc with the inexperienced contractor. Big jobs often include a time-is-of-the-essence clause. Along with this clause is usually a penalty fee that must be paid if the job is not finished on time. The penalty is normally based on a daily fee. For example, you may have to pay $200 per day for every day the job runs past the deadline.

Penalty fees and the possible loss of your bond can ruin your business. Contractors with limited experience in big jobs are often unprepared to project solid completion dates. Don't sign a contract with a completion date that you are not sure you can meet.

COMPETITION IN THE BID PROCESS

When you learn how to eliminate your competition in the bid process, you are on your way to a successful business. There is no shortage of competition in most fields of contracting. There are, however, often shortages of work. With the combination of limited work and unlimited competition, a new business, or any business for that matter, can get discouraged. But don't be—there are ways to thin out the competition.

Beating The Competition With Bid Sheets

Beating the competition with bid sheets is hard. Unless you have a track record and are well known, money will talk. Low prices are what most decision-makers are looking for in bids that are the result of bid sheets. Being the low bidder can get you the job, but you may wish you had never seen the job. Don't bid a job too low. It doesn't do you any good to have work if you're not making money.

How can you improve your odds in mass bidding? If you can get bonded, you have an edge. A lot of bidders can't get bonded. This fact alone can be enough to cull the competition. When you prepare your bid package for submission, be meticulous. All you will have going for you will be your bid package. If you want the job, spend adequate time in preparing a professional bid packet.

In-Person Bids

If you are dealing with in-person bids, follow the guidelines found throughout this book. The basic keys include: dressing appropriately, driving the right vehicle, being professional, being friendly, getting the customer's confidence, producing photos of your work, showing off your letters of reference, giving your bid presentation in person, and following up on all your bids.

PREPARING ACCURATE TAKE-OFFS

To make sound bids, you must be adept at preparing accurate take-offs. It doesn't matter if you use a computerized estimating program or a pen and paper; you must get your facts straight. If you miss items on the take-off, and get the job, you will lose money. If you overestimate the take-off, your price will be too high. An accurate take-off is instrumental in the success of winning a job.

What Is A Take-Off?

A take-off is a list of items needed to do a job. Take-offs are the result of reading blueprints or visiting the job site and making a list of everything you will need to do the job. Some estimators are wizards with take-offs, and others have a hard time trying to project all of their needs. If you can't discipline yourself to learn how to do an accurate take-off, your venture into contracting is going to be a rough road to travel.

Using Take-Off Forms

You can reduce your risk or errors by using a take-off form. If you use a computerized estimating program, the computer files probably already contain forms. You may want to customize the standard computer forms. Whether you are using stock computer forms or making your own forms, you must be sure they are comprehensive.

Take-off forms should have every item you might use in various types of jobs listed. It's best if you create forms that list every expense that you might incur on a job. This improves your odds of reducing omissions when figuring bid prices. In addition, there should be blank spaces on the form that will allow you to fill in specialty items.

The advantage of using take-off forms is that you are prompted on items you might otherwise forget. However, don't get into such a routine that you only look for items on your form. It is very possible a job might require something that you haven't yet put on the form. Forms help, but there is no substitute for thoroughness.

Tracking Inventory

Keeping track of what you've already counted is a problem for some contractors. If you are doing a take-off on a large set of plans, for example a shopping mall, it can be tedious work. The last thing you need to have happen is to lose your place or forget what you've already counted. To avoid this problem, mark each item on the plans as you count it.

Margin Of Error

You should build in a margin of error on your take-off. If you think you are going to need ten rolls of Romex, add a little to your count. How much you add will depend on the size of the job you are figuring. A lot of estimators build in a float figure of between three and five percent. Some contractors add ten percent to their figures. Unless the job is small, I think a ten percent add-on can cause you to lose the bid.

Of course, much of your cushion for mistakes will depend on your ability to make an accurate take-off. If you are good with take-offs, a small percentage for oversights will be sufficient. If you always seem to get on the job and run short of materials, a higher slush pile will be needed.

Keeping Records

It is important that you keep records of your material needs on each job. Don't throw away your take-off. When the job is done, compare the material actually used with what you estimated in your take-off. This will not only help you to see where your money is going, it will make you a better estimator. By tracking your jobs and comparing final counts with original estimates, you can refine your bidding techniques and win more jobs.

PRICING

Pricing your services and materials is an essential part of running a profitable business. If your prices are too low, you may be very busy, but your profits will suffer. If your prices are too high, you will be sitting around, staring at the ceiling, and hoping for the phone to ring. Somewhere between too low and too high is the optimum price for your products and services. The trick is finding out what that price is.

You must learn how to make your prices attractive, without giving away the store. How will you know what price is the right price? Well, you can't pull your prices out of thin air. You must establish your pricing structure with lots of research. When it comes to picking attractive prices, you must look below the surface. There are many factors that can control what you are able to earn. Let's take a look at what is considered a profitable markup.

MARKUP ON MATERIALS

What is a profitable markup on materials? This can be a hard question to answer. It is not difficult to project what a reasonable markup is, but defining a profitable markup is not so easy. Some contractors feel a ten percent markup is adequate. Others try to tack 35 percent onto the price of their materials. Which group is right? Well, you can't make that decision with the limited information I have given you. The contractors that charge a ten percent markup may be doing fine, especially if they deal in big jobs and large amounts of materials. The 35 percent group may be justified in their markup, especially if they are selling small quantities of lower-priced materials.

Markup is a relative concept. Ten percent of $100,000 is much more than 35 percent of $100. For this reason, you cannot blindly pick a percentage of markups to be your firm figure. The figure will need to be adjusted to meet the changes in market conditions and individual job requirements. You can, however, pick percentage numbers for most of your average sales.

If you were in a repair business and were typically selling materials that cost around $20, a 35 percent markup would be fine, if the market would bear it. If you are bidding large housing projects or major commercial jobs, a ten percent markup on materials should be sufficient. To some extent, you have to test the market conditions to determine what price consumers are willing to pay for your materials.

If you are selling common items that anyone can go to the local hardware or building supply and price out, you must be careful not to inflate your prices too much. Customers expect you to mark up your materials, but they don't want to be gouged. If you installed light bulbs with my new light fixture and charged me twice as much for the bulbs as I could have bought them for in the store, I'm not going to be happy. Even though the amount of money involved in the light bulb transaction is puny, the principal of being charged double for a common item still exists.

If you typically install specialty items, you can increase your markup. People will not be as irritated to pay a well marked-up, but reasonable, price for an unusual product. A markup of 20 percent will almost always be acceptable on small residential jobs. When you decide to go above the 20 percent point, do so slowly and test the response of your customers.

MONITORING YOUR COMPETITORS

How can your competitors do work for such low prices? This is a question almost every contractor considers. It always seems to be some company that has a knack for winning bids and beating out the competition. If you know the bids are being won with low prices, you can't help but wonder how the winner of the bids can do it.

Low prices can keep companies busy, but that doesn't mean the low-priced companies are making a profit. Gross sales are important, but net profits are what business is all about. If a company is not making a profit, there is not much sense in operating the business.

Companies that work with low prices fall into several categories. Some companies work on a volume principle. By doing a high volume, the company can make less money on each job and still make a profit by doing many jobs. This type of company is hard to beat.

Some companies sell at low prices out of ignorance. Many small business owners are not aware of the overhead expenses involved with being in business.

This type of inexperienced businessperson will either go out of business quickly or adjust the prices of services and materials. For established contractors competing against newcomers, being able to endure the momentary drop in sales will be enough to weather the storm. In a few months, the new business will either be gone or in a reasonably competitive range.

PRICING YOUR SERVICES

Pricing your services and materials is related directly to your success and longevity. It can be very difficult to decide the right price for your time and material. There are books that give formulas and theories about how to set your prices, but these guides are not always right. Every town and every business will dictate different factors in the prices the public is willing to pay. You can use many methods to find the best fees for your business to charge; let's look at some of them.

Pricing Guides

Pricing guides can be a big help to a business owner who has little knowledge of how to establish the value of labor and materials. However, these guides can cause you frustration and lost business. Most estimating guides provide a formula for adjusting the recommended prices for various regions. The formula used to make this type of adjustment is usually a number that is multiplied against the recommended price. By using the multiplication factor, a price can be derived for services and materials in any major city.

The idea behind these estimating guides is a good one, but there are flaws in the system. I have read and used many of these pricing books. From my personal experience, the books have not been accurate for the type of work I was involved in. I don't mean to say that the books are no good. There are many times when the books are accurate, but I have never been comfortable depending solely on a mass-produced pricing guide.

I have found estimating books to be very helpful as part of the solution to the pricing puzzle. While I don't use pricing figures from these books as my only means of setting a price, I do use them to compare my figures and to ensure I haven't forgotten items or phases of work. Most bookstores carry some form of estimating guides in their inventory.

PRESENTATION

Proper presentation is critical for business success. Even if your price is higher than the competition, you can still win the job with an effective presentation. There are many occasions when the low bidder does not get the job. As a contractor, you can set yourself apart from the crowd by using certain presentation methods. What are these methods? There are numerous ways to achieve an edge over the competition. The ways to win the bid battle can include the way you dress, what you drive, your organizational skills, and much more.

SHOWING YOUR WORTH

When you learn to show a customer why you are worth more money, you are more likely to make the most for your time and effort. People are often willing to pay a higher price to get what they want. You have to convince the customer to want you and your business, not the competition. How can you sway the customers your way? Well, let's study some methods that have proven effective over the years.

Mail Mistakes

You might be surprised how many contractors mail estimates to customers and wait to be called to do the job. Many of these contractors never hear from the customers again. Mailing bids to potential customers is usually not the best way to get the job. When consumers spread estimates out and go over them, it is difficult to see much difference other than price. You want to influence customers with the extras you can bring to the table, so you need a better method of presentation for your quote.

If you must mail your proposal, make sure you prepare a professional package. Use printed forms and stationary, not regular paper with your company name rubber-stamped onto it. Use a heavyweight paper and professional looking colors. Type your estimates and avoid using obvious correction methods to hide your typing errors. If you are mailing your price to the customer, you must make your mailing neat, well organized, attractive, professional, and convincing.

Phone Facts

One of the worst ways to present a formal proposal is by phone. Telephones are great tools for prospecting and following up on estimates, but they are a poor means of delivering initial estimates. When you call in your price, people can't see what you are giving them. They can't linger over the estimate and evaluate it, like they could with a written one. Chances are good that the customer may write down your price and then lose the piece of paper it was written on. Phone estimates also tend to make contractors look lazy, since they don't even take the time to present a professional, written proposal. In general, use the phone to get leads, set appointments, and follow up on estimates. Don't use it to give prices and proposals to possible customers.

Dress Code

The dress code for contractors can span a wide range. Whatever you wear, wear it well. Be neat and clean. Dress in a manner that you can be comfortable with. If you are miserable in a three-piece suit, you will not project as well to the customer. If you normally wear uniforms, you can wear your uniform when presenting your proposal. Jeans are acceptable and so are boots, but both must be clean and neat. Avoid wearing tattered and stained clothing. You don't want the customer to be afraid to ask you sit on the furniture.

When you are deciding on what to wear, consider the type of customer you will be meeting with. If the customer is likely to be dressed casually, then you should dress casually. If you suspect a suit will fit in with your customer's attire, consider wearing a suit. Choose a wardrobe that will blend in with the customer. If you dress too well, you might intimidate the customer. If your clothes and jewelry are too expensive, the customer will probably think you make too much money.

What You Drive

What you drive says a lot about you. If you pull up in front of the average house in a high-priced sports car, you are sending signals to the customer. When the customer looks out the window and sees their contractor getting out of a fancy car, they are going to believe the rumors are true about the outrageous prices contractors charge. If you crawl out of a beat-up, ancient truck, they may assume that you are not very successful.

Choosing the best vehicle for your sales calls is a lot like choosing the proper clothing. You want a vehicle that will make the right statement. A clean van or pick-up should be fine for most any occasion.

Confidence: The Key To Success

Confidence is the key to success. You must be confident in yourself, and you must create confidence in the mind of the consumer. If you can get the customer's confidence, you can almost always get the sale. You will gain your own confidence through experience, but you must learn how to build confidence with your customers.

Gaining the confidence of your customer can often be done by talking. If you are able to sit down with customers and talk for an hour, your odds of getting the job increase greatly. By showing customers examples of your work, you can build confidence. Letters of reference from past customers will help in establishing trust, however, if you have the right personality and sales skills, you can create confidence by simply talking.

As a business owner you are also a sales professional, or at least you had better be. Unless you hire outside sales staff, you are the one customers deal with. If you learn basic sales skills, you will have much more work than the average contractor, even if your prices are higher.

KNOW YOUR COMPETITION

You must know your competition. How you price your services and materials will be affected by the competition. Your prices should be in the same ballpark as your reputable competitors. If your prices are too high, you won't get much work. If your prices are too low, you may be flooded with work and suffer from low profits, not to mention angry competitors.

EFFECTIVE ESTIMATING TECHNIQUES

Effective estimating techniques come in many forms. What works for one person doesn't work for another. By learning from your mistakes and successes, you can mold your own profitable estimating methods. Once you have ways that work, you can use them over and over again. You may have to alter your techniques to match them with specific customers, but the same basic principles that work for you in one sale will work in another.

How do you develop effective estimating techniques? You can learn technical aspects of estimating by reading. Referring to estimating handbooks and pricing guides can perfect your estimating skills further. Much of what you learn will come from experience. Learning from your mistakes can be expensive, but you don't soon forget your costly lessons. One of the most important factors in effective estimating is organization. When you are well organized you are more likely to complete a thorough estimate. Getting work is what it is all about, so go do it!

14

KEEPING JOBS ON TRACK

To move your business along the road to success, you must learn to work with schedules, budgets, and job costing. These three areas of your business can have a strong influence on how much money you make and how long you stay in business. Schedules should be used for many facets of your business. You should have a production schedule, a daily schedule, and a delivery schedule, to name a few. Budgets should also be used for multiple aspects of your business. You should have an overall business budget, an advertising budget, and an inventory budget. Additionally, you will need budgets for individual jobs and the projected growth of your business.

Without the skills to make and follow schedules and budgets, a business owner is likely to make mistakes and lose money. One way to tell if a company is losing money is through the use of job costing. As jobs are in progress, and as they are completed, job costing should be done and monitored. This practice allows the business owner to determine future pricing, forecast cash flow needs, and derive an overall view of the company's on-the-job performance.

Job costing, budgets, and schedules all play a vital role in the development of a contracting business. Without these elements, the business is running in the dark. You can't afford to blindly wander through the business arena. If you do, your competitors will knock you out of the game. Let's see how you can benefit from working intelligently.

ESTABLISHING PRODUCTION SCHEDULES

By establishing production schedules, you can make your business become more organized and more profitable. Production schedules allow you to plan and track your work. This advantage will help you keep track of your workload. Without production schedules, you will have trouble completing your jobs on time. When the jobs run past their completion dates, you will have angry customers on your hands.

How should you go about setting up a production schedule? Working out a viable schedule is not hard and it doesn't take long to do. You may, however, need the help of your subcontractors.

You can rough-up your production schedules on a computer or on notepaper. You will want a separate schedule for each job you have, but start with just one job. Once you have a schedule for the first job, scheduling the remainder of the jobs will be easier. If your jobs will overlap, you may have to schedule multiple jobs at the same time.

Start by putting the job name and address on the schedule. Lay out the schedule with headings for each phase of work you will be responsible for. Create spaces to fill in dates for the various work phases. You will want enough spaces for many dates. The first space will contain your anticipated start date. Another space will contain the date work is actually started. You should have an entry for the estimated time needed to complete the task. Then, you should have a spot to enter the date of actual completion. In addition to your start and finish dates, you should allow spaces to write in dates for material deliveries.

The production schedule should have provisions for listing the names and phone numbers of subcontractors and suppliers you will be depending upon. Write these names and numbers next to each work phase and delivery. Having the names and numbers on the schedule will make it easier to make follow-up confirmation calls.

Now, all you have to do is fill in the appropriate dates. Choosing the dates for your schedule may not be simple. Attempting to work out non-conflicting work schedules can be an arduous task, but it is not impossible. The best way to coordinate your jobs is to talk openly with all the people involved in the jobs.

Call and talk to your subs and suppliers. Go over the first draft of your schedule with them. Ask if they will be available on the dates you want to have them on the job. If you don't know how long it will take the electricians to rough-in the job, ask. When you are using piece workers, you have to treat them like you would any other subcontractor. By following these techniques, the numbers you pencil into the production schedule will be as accurate as you can make them. However, don't expect the dates to remain static. Plans change and so do scheduled dates.

MAINTAINING SCHEDULES

Staying on schedule is going to take effort. You cannot maintain a schedule without working at it. What will you have to do? You will have to make confirmation calls, review the progression of your jobs, stay on top of subs and suppliers, and much more. Let's take a look at some of your scheduling responsibilities.

Confirmation Calls

Making confirmation calls is essential to staying on schedule. If you don't make follow-up calls, subs and suppliers may forget their responsibilities to you. As a contractor, you cannot allow your subcontractors and suppliers to overlook their obligations. It is your job to assemble and manage all the elements of a successful job.

Calling suppliers to confirm orders and deliveries can be done during normal business hours, but this isn't always true with subcontractors. Many subcontractors work in the field. These people don't spend their days sitting next to a phone. To get through to busy subs, you may have to spend time at night making your calls. This can get old, fast.

The nature of your business will determine how much you have to work, but contractors often work from dawn into the late evening. You can reduce the need to work such long hours if you have your subcontractors check in with you at specific times. By asking your subs to call you when they take breaks or lunch, you can eliminate some of the night calls.

Even with the best of efforts, making follow-up calls can become a pain. It may be tempting to let some of your calls slide, but don't do it. If you don't make confirmations on your schedule, the schedule will become useless.

Tracking Production

Tracking production is one of the best ways to keep your schedule accurate. Schedules require adjustments, and to make adjustments, you must keep track of your production. For example, if the drywall contractor is running behind, you will have to reschedule your electricians. If a special-order item, like a custom chandelier, is late coming in, you will have to adjust the schedule. To make these adjustments, you must keep your finger on the pulse of the job. This generally means checking each job's status on a daily basis. After checking your jobs, you can move the schedule up or down to allow for the field production.

Subcontractors and Suppliers

Keeping subcontractors and suppliers in line is another responsibility associated with staying on schedule. If subs or suppliers don't fulfill their obligations in a timely fashion, your schedule will be blown out of the water. You must control your subs and suppliers to stay on schedule.

If you follow my earlier advice and include start and finish dates in your subcontractor agreements, you are on the right track. Include a penalty clause which outlines a daily charge for every day the job is not finished. This gives you an even better chance to stay on schedule. Couple these two contract clauses with persistent prodding, when needed, and you should do fine. If all else fails, exercise your ejection clause and bring in a new sub or supplier to fulfill your needs.

Juggling

There will be times when juggling your schedule will be necessary. While job juggling may be a necessity, it can be very detrimental; be careful. It is one thing to juggle to accommodate unforeseen production changes, but it is quite another thing to juggle for cash flow or greed. Before we go on, let's establish the difference between the two types of job juggling.

Job juggling is necessary and acceptable when materials are late or subcontractors don't perform as expected. This type of juggling is usually associated with extending the estimated completion date, but it might mean moving the completion date up. For example, if you thought it was going to take three days to complete the finish work, and it only took two days, you can move your schedule up by a day. This, of course, is your choice. You could let the existing schedule remain intact.

Juggling your work schedule to generate cash flow or to take on more work is a much more dangerous form of job juggling. If you move your crews from job to job in order to start more jobs, you will probably be out of business within a year. This type of job juggling gives your business a bad reputation and will result in more trouble than it is worth.

UNFORESEEN SCHEDULING OBSTACLES

Adjusting for unforeseen scheduling obstacles gets easier with experience. After being in the business for awhile, you start to anticipate the unexpected. Seasoned contractors are so good at projecting problems that they rarely have them. Unfortunately, contractors that are new to the business suffer through some tough times and pay a price for their lack of experience.

Many people assume that because they know their trade, they know their business. Knowing how to do your job as a tradesperson is not the same as knowing how to run a business. While you might be very competent at estimating the time you will need for your own work, if you're are not an experienced contractor you may not know what allowances to make for subcontractors and suppliers. These two variables can destroy your scheduling plans.

I have hired employees and subcontractors to work with me. I have had some very good workers. During my prime, I rarely found anyone who was substantially faster than I was, but I constantly found workers who were extremely slow. Their slowness was a result of laziness, inexperience, being unfocused, and a number of other reasons. For the same work I can accomplish with a helper in two and a half days, some crews take a full five days. As a contractor, you cannot assume that all subcontractors will complete their tasks in the same periods of time. It is very possible that one sub will need twice as long to get the job done as another. As a contractor, you have to find the secret to meandering through the maze of subcontractors and suppliers.

As you gain experience and get to know your subcontractors and suppliers, you will become more adept at adjusting your schedule. You can almost become a mind reader of how your schedule will need to change. This is not to say that you can avoid all of your scheduling problems, but you can learn to take control of them.

PROJECTING YOUR JOB BUDGET

Projecting your job budget must be one of your priorities. Job budgets are the basics for setting your profit goals. If you miscalculate your job budgets, you will not make the profits you expected to. Occasionally, you may make more money than you expected, but in most cases, you will leave the job with less money than you projected.

For some people the term budget is a dirty word. This mindset extends from personal budgets right into business budgets. While it may not be fun to follow the guidelines of a financial budget, budgets are often the only way to reach your monetary goals.

If you don't project a budget for each of your jobs, you are sure to be disappointed with some of your income results. Budgets may not be enough to keep you from losing money on a job, but they can help you spot your problems and avoid making the same mistake twice.

If you sell a job too cheaply, setting a budget is not going to solve your problem. The best budget going will not compensate for a poorly estimated job. However, by budgeting your jobs, you can evaluate your results and price future jobs to keep your profit margin where you want it, within reason.

How should you structure a job budget? To make a job budget, you must project and list all anticipated costs of the job. For some contractors, thinking of all the expenses to be incurred in a job is a chore they cannot fathom. These people don't have the natural ability to put all steps of a job into chronological order. This doesn't mean the contractors can't learn to forecast a budget; they just are not naturally talented in this aspect of running a business. Building a job budget is not difficult, if you proceed in a logical way. Let's see how you can structure a stable job budget.

Like building a production schedule, label your form with the job name and address. Continue by listing all aspects of the work phases. For example, if you are going to forecast your expenses, include spaces for ground works, rough-ins, fixtures, trim-out work, and so forth.

Most contractors don't have problems in listing the large phases of work. However, many business owners neglect to list the smaller phases of work required on a job. These minor phases can add up to quite a lot of money.

Consider all of your anticipated costs. Let's look at some of the costs that often go undetected in a budget. These costs could include the following:

- *Design drawings*
- *Permit fees*
- *Trash and disposal fees*
- *Clean-up costs*
- *Administration costs*
- *Office overhead expenses*
- *Inspection fees*
- *Field supervisors*
- *Advertising costs incurred to obtain the job*
- *Time spent estimating and selling the job*
- *Potential warranty work*

There are many costs associated with doing a job that will not show up on most contractors' budgets. If these soft-costs are left out, the profits from the job will shrink. You must be thorough when itemizing your job expenses. If you are not, the budget is bogus.

PROJECTING YOUR BUSINESS BUDGET

Projecting your business budget is similar to forecasting a job budget. In both instances, you must account for all anticipated expenses. Unlike a job budget, a business budget will have to include projections for a much longer period of time. This extended time can make the creation of a good business budget more difficult to master.

Developing a business budget that will work requires extensive thought. You must consider your immediate needs and your projected future needs. Your desires will come into play with a business budget. For example, if you want to retire in twenty years, you must make an allowance for this desire in your budget. Let's take a look at some of the factors you must incorporate into the budget for your business.

Salary

Many business owners don't pay themselves a set salary. They make as much money as they can and use the money they need. This works for some people, but it is not the way to make a business budget.

When casting your business budget, you need to establish a set amount for your salary. You may not always be able to take as much money for your salary as the budget reflects, but you need a number plugged in for your income. When you set your salary, don't forget your responsibility for income taxes. You have to look at your net income needs and the gross income requirements. There is a substantial difference between your net needs and your gross needs.

Office Expenses

Office expenses can cause your budget to balloon. Think about all the various expenses that you incur to keep your office running. These expenses may include: rent, utilities, phone bills, cleaning, equipment rentals, office supplies, furniture, and much more. The costs of your office expenses can account for a high percentage of your annual expenses. Many contractors have excessive office expenses. When you build your business budget, check over your office expenses; you will likely find ways to save money.

Field Expenses

For a number of contractors, field expenses are a large part of the annual budget. Field expenses can be broken down into more specific categories. These costs include: vehicles, fuel, field supervisors, signs, and other related expenses.

Vehicle Expenses

Almost every contracting business is affected by the expense of owning and maintaining vehicles. The cost of trucks, cars, fuel, tires, and similar expenses can amount to thousands of dollars. Of course, your vehicle expenses will be related directly to the size and structure of your business. If you rely on subcontractors for all of your fieldwork, your vehicle expenses will be minimal. However, if you have a fleet of trucks and an army of employees, vehicle expenses can be astronomical.

Tool and Equipment Expenses

There are two types of tool and equipment expenses. There is the cost of initial acquisitions and the cost of replacement. Even if you have all the tools and equipment you need, you have to budget for replacement costs. Your hardware may be broken, stolen, or worn out, but sooner or later, it will need to be replaced. If you are not prepared financially for these replacements, you can find yourself unable to continue doing business. Take a look at your tools and equipment. Estimate the life expectancy of your hardware and put a figure in your budget for its replacement.

Employee Expenses

Employee expenses can be devastating. When you look at employee expenses, you have to look much deeper than the hourly rates earned by your employees. You must look at the taxes involved with having employees. Other considerations are employee benefits. If you provide insurance benefits for your employees, you are spending some serious money. Paid vacations are another major expense. Sick leave, paid holidays, and other similar employee benefits can amount to thousands of dollars quickly.

If your business is top heavy with employee expenses, you might do well to consider engaging independent contractors. When employees are involved, don't neglect to include the cost of employee benefits in your budget.

Insurance Expenses

Insurance expenses can run some businesses into the ground. Insurance needs vary, depending on the nature of various businesses. Liability insurance is a need for all businesses. When employees are used, workman's compensation insurance can become an expensive factor in the business budget. Insurance to protect against theft and fire is another common business need. When a company is properly insured, the expense of the insurance can account for much of the total business budget.

Advertising Expenses

Advertising expenses are one of the few expenses that businesses can directly profit from. When used efficiently, advertising will pay for itself and then some. However, a business owner can never be sure how effective advertising will be.

How much will your advertising needs amount to? While you cannot be sure what your advertising results will be, you can project how much you are willing to spend on generating new business. Many companies dedicate a percentage of their gross sales to advertising. Most small companies pick a dollar amount, rather than a percentage. Whichever method you use, make sure to include your anticipated advertising costs in your business budget.

Loan Expenses

Loan expenses can be overlooked when building a budget. This is especially true if you will use short term, interest-only loans. The cost of these loans may not register as an expense. The reason is that the interest is usually paid in a lump sum, not on a monthly basis. Don't forget to include the fees you will incur with your business financing.

Taxes

Taxes are another expense that generally is left out of a business budget, but it shouldn't be. The impact of taxes can cause severe stress on a business. Unless regular tax deposits are made, the burden of coming up with enough money to satisfy the tax authorities can be overwhelming.

To avoid being caught in a cash bind, include your estimated tax liabilities on your budget projections. If you are unable to forecast your taxes, consult a tax professional; you cannot afford to be left high and dry on tax day.

Growth Expenses

To see business growth, you must plan for growth expenses. Few businesses grow without a plan. Part of any growth plan will involve expansion capital. Consider your intent for company growth when building your budget, and include provisions for the cost of expansion.

Retirement Goals

Retirement goals frequently get pushed to the back of the list. When money is tight, it is hard to invest in retirement plans. The impulse is to fight today's fires and worry about retirement later. Unfortunately, age and retirement creeps up on all of us, often sooner than we planned for.

New business owners often see the business they are building as their retirement. In some cases their beliefs are founded, but more often, the business is not enough for retirement. In fact, many of the new businesses won't be in operation at the time of the owner's desire to retire. For these reasons, investments in your business are not enough for safe retirement planning.

To build a suitable retirement portfolio, you need a plan. The plan will normally involve diversified investments. Examples could include: stocks, bonds, real estate, coins, art, and a number of other possibilities. It is a wise idea for most people to talk with professionals to obtain a viable retirement plan. Once you have a solid path to retirement, include the needs for funding your retirement in the business budget.

As you build your business budget, you may discover other items to account for, but this list of routine expenses should get you started in the right direction. Remember, you wouldn't take a trip to a strange destination without directions, and your budget is to your business what a map is to your travel plans.

MAINTAINING FINANCIAL BUDGET REQUIREMENTS

Having a budget will do little good unless you have the discipline and skills for maintaining your financial budget requirements. It helps to have a natural ability to maintain budget constraints, but if you don't, you can learn the skills and discipline.

Self-discipline is a factor throughout our lives. It is this quality that allows us to live among the laws of society. Some people have strong discipline towards certain aspects of life and no discipline in others. Compulsive gamblers may discipline themselves to going to work on time and maintaining an average life, until it comes to gambling. People with an obsession for food may be perfectly normal, until food is set in front of them. Most of us have some weak areas in our self-discipline. Your job as a business owner is to find your weak areas and reinforce them.

If you find the deals on tools in mail-order catalogs irresistible, you may be spending far too much on tools. When you are infatuated with having the most high tech office equipment available, you may be investing too much money on the wrong items. The list of potential weaknesses could go on, but you get the idea.

All of these potential weaknesses must be evaluated. If you are quick to give in to impulse buying, you must set rules for yourself.

Make yourself wait for a reasonable period of time before buying that new tool or piece of office equipment. Wait and see if you need the item or if you only want the item. Look into every aspect of your business habits. Are you going to lunch at expensive restaurants on a regular basis? If you are, you may be spending enough to pay for a more justified business expense.

Document all of your spending for the next two weeks; include every item you spend a dime on. At the end of the two weeks, go over your list of expenditures. Compare these expenses with your budget. How many of your recent expenditures are not reflected in your budget? My guess is that you will find several areas where you are spending money that wasn't budgeted. This is a danger signal; you can't allow yourself to run rampant with your company's financial resources. If you do, the business will fail.

After tagging all the unbudgeted expenses, decide if you will continue to make similar purchases. If you come to the conclusion that your expenses are justified, include them in the budget. If the spending habits are not necessary, eliminate them. Repeat the two-week test periodically. By monitoring and refining your budget, you can stay within its framework.

JOB COSTING: ASSESSING FUTURE JOB PRICING

Job costing is your way of assessing future job pricing. To make the most of your business, you must obtain the best price possible for your services. Sometimes this will mean increasing your prices, and sometimes it will mean lowering your prices. How will you know which way to move your price pointer? You can develop a strong sense of direction by reviewing and assessing your profit performance on past jobs. Job costing will give you the information you need to make sound decisions on your pricing structure.

Job costing is the act of adding up all the costs incurred to complete a job. Depending upon your methods, job costing may only account for labor and materials used on the job. In more sophisticated reports, the numbers will include the cost of overhead and operation expenses.

Your routine business expenses must be factored into the cost of jobs, but you don't have to include it in individual job-cost reports. You can, instead, use a percentage of your on-job profit to defer your other costs of doing business. Let's see how each of these methods can be used.

Simple Job Costing

Simple job costing will include all expenses incurred for a specific job, excluding overhead expenses. Overhead expenses might include: office rent and utilities, general insurance, advertising, and so forth.

To do a simple job cost, you need nothing more than pencil and paper, and information. First, list all materials used on the job. Then, price these materials based on what you paid for them. Next, calculate all the labor that went into the job. Once you have all the labor accounted for, give it a value. The value should be your cost for the labor. Soft costs will be your next category. Soft costs could include permit fees, the cost of blueprints, and so on.

Once you have all the costs for the job listed, check over the list for omissions. When you are sure you haven't forgotten anything, add up the total of the costs. Subtract the total of your costs from the money you received for doing the job, and you've got your gross profit.

In real life, it is not so simple to account for overhead expenses. It is not uncommon to have several jobs running at once and many jobs overlapping each other. This complicates the division of overhead between jobs. Some businesses project a percentage of their overhead and apply it to every job. Other businesses use complex methods to detail the exact cost of their overhead to each job. How you do your job costing is up to you, but you must allow for the off-job expenses to get a true impression of your profit margin. Now, let's see how a job costing report might be done using a blanket percentage for overhead expenses.

Percentage Job Costing

Percentage job costing is another common method for determining profits and losses. To perform this task, duplicate the instructions given for simple job costing. Once this is done, subtract a percentage of the contract price for the job. The percentage you subtract will represent an estimate of all your overhead expenses.

What percentage should you use? The percentage applied for overhead will depend on your business structure. Some businesses carry heavy overhead expenses and others are streamlined to maximum efficiency. You will have to experiment with your personal circumstances to determine what percentage will cover your off-job expenses.

Adjusting your prices will be much easier when you are working with the results of accurate job costing reports. After reviewing the job costs on three similar jobs, you may see a pattern. If so, you can determine if you are charging enough, too much, or not enough.

TRACKING PROFITABILITY

Tracking your profitability is possible from job costing, but there is much more than can be accomplished. Job costing allows you to monitor all of your jobs for maximum profit potential. It is easy to say you want to make a gross profit of 20 percent on every job, but it is not so easy to accomplish. Some jobs yield a higher profit potential than others. This is where job costing can lead you in the right direction.

To see how different jobs produce various results, let's consider the circumstances of a general contractor. This contractor deals only in residential jobs. The jobs range from minor remodeling to major renovations. New construction, such as houses, garages, and additions, also makes up a part of this business.

The general contractor has allocated $15,000 for advertising over the next year. Knowing how much he has to spend on advertising, the contractor must decide what to spend it on. One consideration is the type of advertising media to use, but beyond that is the question of what types of jobs will produce the most profits.

By going over past job cost reports, the contractor can determine what types of jobs offer the most profit potential. In reviewing past performances, the contractor may find that attic conversions are the most profitable jobs to undertake. It might be discovered that building garages is a quick and easy source of substantial profits. Perhaps the construction of additions will provide the most net income. Building houses yields the highest overall income, but it can take a long time for all of the money to be earned.

There are some industry standards for which jobs should produce the most money, but all business owners must identify what jobs offer the most benefits to their companies. While building decks might make the most economic sense for Larry and Ann, Joe may find his highest profits come from kitchen and bathroom remodeling. Your big moneymaker might be townhouse projects. Maybe your profits are highest in service and repair work. Remodeling work is usually very profitable for plumbers and electricians, so maybe this is your niche. What types of electrical work do you see the highest profits from?

The degree of profits from various jobs will depend on the companies performing them. Often, companies are more efficient at some jobs than others. For your company, you can look over the results of your completed projects to forecast your future.

Job costing does much more than just tell you how much you made or lost. It allows you to outline a plan for your business. Advertising is more effective when it is the result of carefully studying job cost reports. Budget distributions are easier to establish when you have detailed job costs to pull from. There is almost no limit to how accurate job cost reports can help your business.

ACCURATE JOB COSTING

Accurate job costing is essential to long-term success. To endure the test of time, businesses must be financially sound and flexible. One reason so many new businesses fail is the inexperience of their owners and operators. These people may have solid work skills, but lack business skills. It is not uncommon to find first-time contractors working hard and believing they are making money, only to find at the end of the year that they would have made more as an employee than as an owner. These inexperienced business owners are not aware of the time, effort, and skills needed to guide a business to success. Many of them think that making $45 per hour is fantastic. Even when these hourly earnings are double what the individual was making as an employee, the end result may be lower net pay.

Because rookie contractors never consider so many expenses, they go about their business with stars in their eyes. Then, one day, the clouds move in and cover the stars. This is the day when the new contractor realizes how much money is needed to keep the business operating. All of a sudden, that $45 per hour is nowhere near enough to cover the business and personal needs.

You can avoid this trap by building a budget for all the businesses expenses you will incur and maintaining that budget. Job costing will show you where you stand financially. You will discover categories that were left out of your budget. Job cost reports can bring the reality of overhead expenses to light. While you may have never considered charging a portion of your rent to a job, you will see that you must.

The information you gain from accurate job costs will make you a better estimator. If you lost money or made a minimal profit on your last job, the next job should go better. Since you will be able to pinpoint your problems with job costing, you can adjust your next estimate to protect yourself.

One area that causes a lot of problems for contractors is travel time. Many business owners underestimate the amount of time that will be spent on the road. Job cost reports can open your eyes to mistakes in this category, too.

Since the labor on your job costs will be broken down into phases, you will be able to see how time was spent. One problem area to look for is time wasted on trips to the supply house. This is a common robber of profits for many electricians.

TIME MANAGEMENT

If you can eliminate lost time running to the supply house, you will make more money. Wasted time on the road is one of the most prevalent causes of lost income to contractors. It may not be one of the larger financial losses a company will experience, but it is often the most regular loss of income. Electricians who are not able to make accurate take-offs and schedule deliveries properly lose money on the road. When they or their employees leave the job for missing materials, profits are eroding.

During my time as a contractor and a consultant, I have witnessed countless situations when runs to the supply house pulled the job profits down. Most of these trips were avoidable, but were made anyway. Even circumstances that are not reasonably avoidable can be made better with the use of logic. As a business owner, you cannot afford this type of wasted time. Except in rare cases, there is no suitable excuse for it. If you spend time preparing for a job, you will not waste time running for materials. To run profitable jobs, you have to make every effort to do so. You might not like paperwork, but if you enjoy seeing a large bank balance in your account, you had better be willing to take care of business matters as well as you install electrical systems.

15

WORKSITE SAFETY

Worksite safety doesn't get as much attention as it should. Far too many people are injured on jobs every year. Most of the injuries could be prevented, but they are not. Why is this? People are in a hurry to make a few extra bucks, so they cut corners. This happens with electrical contractors and piece workers. It even affects hourly electricians who want to shave 15 minutes off their workday, so that they can head back to the shop early.

Based on my field experience, most accidents occur as a result of negligence. Electricians try to cut a corner, and they wind up getting hurt. This has proved true with my personal injuries. I've only suffered two serious on-the-job injuries, and both of them were a direct result of my carelessness. I knew better than to do what I was doing when I got hurt, but I did it anyway. Well, sometimes you don't get a second chance, and the life you affect may not be your own. So, let's look at some sensible safety procedures that you can implement in your daily activities.

FACING DANGERS

Electrical work can be a very dangerous trade. The tools of the trade have the potential to be killers. Requirements of the job can place you in positions where a lack of concentration could result in serious injury or death. The fact that electrical work can be dangerous is no reason to rule out the trade as your profession. Driving can be extremely dangerous, but few people refuse to set foot in an automobile out of fear.

Fear is generally a result of ignorance. When you have a depth of knowledge and skill, fear begins to subside. As you become more accomplished at what you do, fear is forgotten. While it is advisable to learn to work without fear, you should never work without respect. There is a huge difference between fear and respect.

If, as an electrician, you are afraid to climb up on a roof to check a weather head, you are not going to last long in the trade. However, if you scurry up the roof recklessly, you could be injured severely, perhaps even killed. You must respect the position you are putting yourself in. If you are using a ladder to get on the roof, you must respect the possible outcome of a mistake. Once you are on the roof, you must be conscious of footing conditions, and the way that you negotiate the pitch of the roof. If you pay attention, are properly trained, and don't get careless, you are not likely to get hurt.

Being afraid of a roof will limit or eliminate your electrical career. Treating your trip to and from the roof like a walk into your living room could be deadly. Respect is the key. If you respect the consequences of your actions and you are aware of what you are doing, the odds for a safe trip improve.

Many young electricians are fearless in the beginning. They think nothing of darting around on a roof or jumping down in a trench. As their careers progress, they usually hear about or see on-the-job accidents. Someone gets buried in the cave-in of a trench. Somebody falls off a roof. A metal ladder hits a power line as it is being set up. A careless electrician steps into a flooded basement and is electrocuted because of submerged equipment. The list of possible job-related injuries is a long one.

Millions of people are hurt every year in job-related accidents. Most of these people were not following solid safety procedures. Sure, some of them were victims of unavoidable accidents, but most were hurt by their own hand, in one way or another. You don't have to be one of these statistics.

In over 25 years in the trades, I have only been seriously hurt on the job twice. Both times were my fault. I got careless. In one of the instances, I let loose clothing and a powerful drill work together to chew up my arm. In the other incident, I tried to save myself the trouble of repositioning my stepladder while drilling holes in floor joists. My desire to save a minute cost me torn stomach muscles and months of pain.

My accidents were not mistakes; they were stupidity. Mistakes are made through ignorance. I wasn't ignorant of what could happen to me. I knew the risk I was taking, and I knew the proper way to perform my job. Even with my knowledge, I slipped up and got hurt. Luckily, both of my injuries healed, and I didn't pay a lifelong price for my stupidity.

During my long career I have seen a lot of people get hurt. Most of these people have been helpers and apprentices. Of all the on-the-job accidents I have witnessed, every one of them could have been avoided. Many of the incidents were not extremely serious, but a few were.

As an electrician, you will be doing some dangerous work. You will be drilling holes, running threading machines, and a lot of other potentially dangerous jobs. Hopefully, your employer will provide you with quality tools and equipment. If you have the right tool for the job, you are off to a good start in staying safe.

Safety training is another factor you should seek from your employer. Some contractors fail to tell their employees how to do their jobs safely. It is easy for someone, like an experienced electrician who knows a job inside and out, to forget to warn an inexperienced person of potential danger.

For example, a seasoned electrician might tell you to break up the concrete around a conduit and never consider telling you to wear safety glasses. The electrician will assume you know the concrete is going to fly in your face as it is chiseled up. However, as a rookie, you might not know about the reaction concrete has when hit with a cold chisel. One swing of the hammer could cause extreme damage to your eyesight.

Simple jobs, like the one in the example, are all it takes to ruin a career. You might be really on your toes when asked to scoot across an I-beam, but how much thought are you going to give to tightening the bolts on a fixture? The risk of falling off the I-beam is obvious. Having chips of glass from a broken fixture flying into your eyes is not so obvious. Either way, you can have a work-stopping injury.

Safety is a serious issue. Some job sites are very strict in the safety requirements maintained. But a lot of jobs have no written rules of safety. If you are working on a commercial job, supervisors are likely to make sure you abide by the rules of the Occupational Safety and Health Administration (OSHA). Failure to comply with OSHA regulations can result in stiff financial penalties. However, if you are working on residential jobs, you may never set foot on a job where OSHA regulations are observed.

In all cases, you are responsible for your own safety. Your employer and OSHA can help you to remain safe, but in the end, it is up to you. You are the one who has to know what to do and how to do it. You do not only have to take responsibility for your own actions, you also have to watch out for the actions of others. It is not unlikely that you could be injured by someone else's carelessness. Now that you have had the primer course, let's get down to the specifics of job-related safety.

As we move into specifics, you will find the suggestions in this chapter broken down into various categories. Each category will deal with specific safety issues related to the category. For example, in the section on tool safety, you will learn procedures for working safely with tools. As you move from section to section, you may notice some overlapping of safety tips. For example, in the section on general safety, you will see that it is wise to work without wearing jewelry. Then you will find jewelry mentioned again in the tool section. The duplication is done to pinpoint definite safety risks and procedures. We will start into the various sections with general safety.

GENERAL SAFETY

General safety covers a lot of territory. It starts from the time you get into the company vehicle and carries you right through to the end of the day. Much of the general safety recommendations involve the use of common sense (Figure 15.1). Now, let's get started.

Vehicles

Many electricians are given company trucks for their use in getting to and from jobs. You will probably spend a lot of time loading and unloading company trucks. And, of course, you will spend time either riding in or driving them. All of these areas can threaten your safety.

If you will be driving the truck, take the time to get used to how it handles. Loaded work trucks don't drive like the family car. Remember to check the vehicle's fluids, tires, lights, and related equipment. Many work trucks are old and have seen better days. Failure to check the vehicles' equipment could result in unwanted headaches. Also, remember to use the seat belts; they do save lives.

Apprentices are normally charged with the duty of unloading the truck at the job site. There are a lot of ways to get hurt in doing this job. Many trucks use roof racks to haul conduit and ladders. If you

General Safe Working Habits

1. Wear safety equipment.

2. Observe all safety rules at the particular location.

3. Be aware of any potential dangers in the specific situation.

4. Keep tools in good condition.

FIGURE 15.1 General safe working habits.

are unloading these items, make sure they will not come into contact with low-hanging electrical wires. If you are unloading heavy items, don't put your body in awkward positions. Learn the proper ways of lifting, and never lift objects inappropriately. If the weather is wet, be careful climbing on the truck. Step bumpers get slippery, and a fall can impale you on an object or bang up your knee.

When it is time to load the truck, observe the same safety precautions you did in unloading. In addition to these considerations, always make sure your load is packed evenly and well secured. Be especially careful of any load you attach to the roof rack, and always double check the cargo doors on trucks with utility bodies.

CLOTHING

Clothing is responsible for a lot of on-the-job injuries (15.2). Sometimes it is the lack of clothing that causes the accidents, and there are many times when too much clothing creates the problem. Generally, it is wise not to wear loose fitting clothes. Shirttails should be tucked in, and short-sleeve shirts are safer than long-sleeved shirts when operating some types of equipment.

Caps can save you from minor inconveniences, like getting cobwebs in your hair, and hard hats provide some protection from potentially damaging accidents, like having a steel fitting dropped on your head. If you have long hair, keep it up and under a hat.

Good footwear is essential in the trade. Normally a strong pair of hunting-style boots will be best. The thick soles provide some protection from nails and other sharp objects you may step on. Boots

---

Safe Dressing Habits

1. Do not wear clothing that can be ignited easily.

2. Do not wear loose clothing, wide sleeves, ties or jewelry (bracelets, necklaces) that can become caught in a tool or otherwise interfere with work. This caution is especially important when working with electrical machinery.

3. Wear gloves to handle hot or cold pipes and fittings.

4. Wear heavy-duty boots. Avoid wearing sneakers on the job. Nails can easily penetrate sneakers and cause a serious injury (especially if the nail is rusty).

5. Always tighten shoelaces. Loose shoelaces can easily cause you to fall, possibly leading to injury to yourself or other workers.

6. Wear a hard hat on major construction sites to protect the head from falling objects.

FIGURE 15.2   Safe dressing habits.

with steel toes can make a big difference in your physical well-being. If you are going to be climbing, wear footgear with a flexible sole that grips well. Gloves can keep your hands warm and clean, but they can also contribute to serious accidents. Wear gloves sparingly, depending upon the job you are doing. Long sleeved shirts will protect your arms from hot metal particles when you are drilling, grinding, or threading metal.

JEWELRY

On the whole, jewelry should not be worn in the workplace. Rings can inflict deep cuts in your fingers. They can also work with machinery to amputate fingers. Chains and bracelets are equally dangerous, probably more so. Arguably, the most dangerous piece of jewelry an electrician can wear to work is a wristwatch with a metal strap. This can be very hazardous when working around energized parts. It is wise to wear a watch with a non-metallic strap, and it is strongly recommended that you carry your watch in your pocket when you are at work.

Trade Tip: Keeping your tools in good condition is a wonderful way to make your work more productive, more profitable, and safer. Keep drill bits and saw blades sharp. Check electrical cords every day to see that they are safe. Inspect your equipment and use it for the purpose it is intended to be used for.

EYE AND EAR PROTECTION

Eye and ear protection is often overlooked. An inexpensive pair of safety glasses can prevent you from spending the rest of your life blind. Ear protection reduces the effect of loud noises, such as jackhammers and drills. You may not notice much benefit now, but in later years you will be glad you wore it. If you don't want to lose your hearing, wear ear protection when subjected to loud noises.

PADS

Kneepads not only make an electrician's job more comfortable, they help to protect the knees. Some electricians spend a lot of time on their knees, and pads should be worn to ensure that they can continue to work for many years.

The embarrassment factor plays a significant role in job-related injuries. People, especially young people, feel the need to fit in and to make a name for themselves. Electrical work is sort of a macho trade. There is no secret that electricians often fancy themselves as strong human specimens. Many electricians are strong. The work can be hard and, in doing it, becoming strong is a side benefit. But

Don't Do This! Don't wear sneakers on a job site. Boots with heavy soles provide a lot more protection from nails and other sharp items that can injure your feet. I see people with sneakers constantly. They may be fashionable, but they are not practical for most job-site applications.

you can't allow safety to be pushed aside for the purpose of making a macho statement.

All too many people believe that working without safety glasses, ear protection, and so forth makes them tough. That's just not true. It may make them appear dumb and it may get them hurt, but it does not make them look tough. If anything, it makes them look stupid or inexperienced.

Don't fall into the trap so many young electricians do. Never let people goad you into bad safety practices. Some workers are going to laugh at your kneepads. Let them laugh; you will still have good knees when they are hobbling around on canes. I'm dead serious about this issue. There is nothing sissy about safety. Wear your gear in confidence, and don't let the few jokesters get to you. I use a dense foam pad to kneel on when I'm installing receptacles or any other action requiring kneeling. One of the carpenters I work with frequently calls it my "girly-pad." Privately, he sheepishly admits that it is an extremely prudent safety measure.

TOOL SAFETY

Tool safety is a big issue. Anyone in the trade will work with numerous tools. All of these tools are potentially dangerous, but some of them are especially hazardous. This section is broken down by the various tools used on the job. You cannot afford to start working without the basics in tool safety. The more you can absorb on tool safety, the better off you will be (Figure 15.3).

The best starting point is reading all the literature available from the tool manufacturers. The people that make the tools provide some good safety suggestions. Read and follow the manufacturers' recommendations.The next step in working safely with your tools is to ask questions. If you don't understand how a tool operates, ask someone to explain it to you. Don't experiment on your own; the price you pay could be much too high.

Common sense is irreplaceable in the safe operation of tools. If you see an electrical cord with cut insulation, you should have enough common sense to avoid using it. In addition to this type of simple observation, you will learn some interesting facts about tool safety. Now, let me tell you what I've learned about tool safety over the years.

There are some basic principles to apply to all of your work with tools. We will start with the basics, and then we will move on to specific tools. Here are the basics:

- *Keep body parts away from moving parts.*
- *Don't work with poor lighting conditions.*
- *Be careful of wet areas when working with electrical tools.*
- *Always plug power tools into a GFCI protected source.*
- *If special clothing is recommended for working with your tools, wear it.*
- *Use tools only for their intended purposes.*
- *Always grip power tools firmly with both hands.*
- *Get to know your tools well.*
- *Keep your tools in good condition.*

Now, let's take a close look at the tools you are likely to use. Electricians use a wide variety of hand tools and electrical tools. They also use specialty tools. So let's see how you can use all these tools without injury (Figure 15.4).

Drills And Bits

Drills have been my worst enemy. The two serious injuries I have received were both related to my work with a right-angle drill. The

---

Safe Use of Hand Tools

1. Use the right tool for the job.

2. Read any instructions that come with the tool unless you are thoroughly familiar with its use.

3. Wipe and clean all tools after each use. If any other cleaning is necessary, do it periodically.

4. Keep tools in good condition. Chisels should be kept sharp and any mushroomed heads kept ground smooth; saw blades should be kept sharp; pipe wrenches should be kept free of debris and the teeth kept clean; etc.

5. Do not carry small tools in your pocket, especially when working on a ladder or scaffolding. If you should fall, the tools might penetrate your body and cause serious injury.

FIGURE 15.3 Safe use of hand tools.

drills used most by electricians are not the little pistol-grip, hand-held types of drills most people think of. The day-to-day drilling done by electricians involves the use of large, powerful drills. These drills have enormous power when they get in a bind. Hitting a nail or a knot in the wood being drilled can do a lot of damage. You can break fingers, lose teeth, suffer head injuries, and a lot more. As with all electrical tools, you should always check the electrical cord before using your drill. If the cord is not in good shape, don't use the drill.

Always know what you are drilling into. If you are doing new construction work, it is fairly easy to look before you drill. However, drilling in a remodeling job can be much more difficult. You cannot always see what you are getting into. If you are unfortunate enough to drill into a hot wire, you can get a considerable electrical shock. It is always wise to look twice and drill once, rather than looking quickly and drilling only to damage existing wiring.

The bits you use in a drill are part of the safe operation of the tool. If your drill bits are dull, sharpen them. Dull bits are much more dangerous than sharp ones. When you are using a standard ship auger

Safe Use of Electric Tools

1. Always use a three-prong plug with an electric tool.

2. Read all instructions concerning the use of the tool (unless you are thoroughly familiar with its use).

3. Make sure that all electrical equipment is properly grounded. Ground fault circuit interrupters (GFCI) are required by OSHA regulations in many situations.

4. Use proper-sized extension cords. (Undersized wires can burn out a motor, cause damage to the equipment, and present a hazardous situation.

5. Never run an extension cord through water or through any area where it can be cut, kinked, or run over by machinery.

6. Always hook up an extension cord to the equipment and then plug it into the main electrical outlet—not vice versa.

7. Coil up and store extension cords in a dry area.

FIGURE 15.4 Safe use of electric tools.

Trade Tip: Don't leave electrical tools connected to a power source when you are not using them. Imagine a scene where a homeowner comes to a job to inspect the progress of the job. Assume that this homeowner has a young child along for the visit. If the child picks up a circular saw that is connected to electrical power and pulls the trigger, the results may not be something that you would want to live with for the rest of your life. Safety on the job is not only about you.

bit to drill through thin wood, like plywood, be careful. Once the worm driver of the bit penetrates the plywood fully, the big teeth on the bit can bite and jump, causing you to lose control of the drill. If you will be drilling metal, be aware that the metal shavings will be sharp and hot. Be careful when your bit punches through the material being drilled. The sudden lunge forward can rake your hands across objects, leaving you with a nasty gash in your hand or wrist.

Power Saws

Electricians don't use power saws as much as carpenters, but they do use them. The most common type of power saw used by electricians is the reciprocating saw. These saws are used to cut pipe, plywood, floor joists, and a whole lot more. In addition to reciprocating saws, electricians use circular saws and chop saws. All of the saws have the potential for serious injury.

Reciprocating saws are reasonably safe. Most models are insulated to help avoid electrical shocks if a hot wire is cut. The blade is typically a safe distance from the user, and the saws are pretty easy to hold and control. However, the brittle blades do break. This could result in an eye injury. If you loose control of a reciprocating saw and it comes out of the material being cut, it can jackhammer across the surface. Never place body parts on the surface of the material being cut.

Circular saws are used occasionally. The blades on these saws can bind and cause the saws to kick back. Chop saws are sometimes used to cut pipe. If you keep your body parts out of the way and wear eye protection, chop saws are not unusually dangerous.

Don't Do This! Don't use tools for anything that they are not intended to be used for. As an example, don't use a screwdriver as a chisel or as a pry tool. Many injuries occur when tools are used for purposes other than for what they were designed.

Power Pipe Threaders

Power pipe threaders are very nice to have if you are doing much work with threaded pipe. However, these threading machines can grind body parts as well as they thread pipe. Electric threaders are very dangerous in the hands of untrained people. It is critical to keep fingers and clothing away from the power mechanisms. The metal shavings produced by pipe threaders can be very sharp, and burrs left on the threaded pipe can slash your skin. The cutting oil used to keep the dies from getting too hot can make the floor around the machine slippery.

Occasionally, electricians need to insert bent pieces of conduit into a threading machine. Care must be taken to avoid being caught by the rotating conduit and to insure that the bent rotating piece will clear the floor and walls of the threading area. Always place a piece of plywood or cardboard underneath the threading machine to catch spatters of threading oil. Ideally a sheet of plywood with a two by four rim around it, filled with two or three inches of sand, should be set up beneath the threading machine. This application is more suited for larger jobs.

Air-Powered Tools

Electricians do not often use air-powered tools. Jackhammers are probably the air-powered tools used most by electricians. When using tools with air hoses, check all connections carefully. If you experience a blowout, the hose can spiral wildly out of control. Sometimes electricians use compressed air to blow pull lines into conduit ducts. Eye protection should be worn to prevent foreign objects from blowing back into your eyes.

Powder-Actuated Tools

Powder-actuated tools are used by electricians to secure objects to hard surfaces like concrete. If the user is properly trained, these tools are not too dangerous. However, good training, eye protection, and ear protection are all necessary. Misfires and chipping hard surfaces are the most common problems with these tools. Powder-actuated tool manufacturers are usually more than happy to train and license users of their equipment.

Ladders

Electricians use stepladders and extension ladders frequently. Many ladder accidents are possible. You must always be aware of what is around you when handling a ladder. Ladders used by electricians must always be constructed of a nonconductive material such as fiberglass or wood. Metal ladders should never be used because if you brush against a live electrical wire with a ladder you are carrying, your life could be over. Ladders often fall over when the people using them are not careful. Reaching too far from a ladder can be all it takes to cause you to fall (Figure 15.5).

When you set up a ladder or rolling scaffolds, make sure it is set up properly. The ladder should be on firm footing, and all safety braces and clamps should be in place. When using an extension ladder, many electricians use a rope to tie rungs together where the sections overlap. The rope provides an extra guard against the ladder's safety clamps failing and the ladder collapsing. When using an extension ladder, be sure to secure both the base and the top. I had an unusual accident on a ladder that I would like to share with you.

I was on a tall extension ladder, working near the top of a commercial building. The top of my ladder was resting on the edge of the flat roof. There was metal flashing surrounding the edge of the

 Don't Do This! Don't leave tools perched on the top of tall ladders. Even an 8 foot stepladder is tall enough for a falling tool to do some serious damage to a human body. It's far too easy for a ladder to be bumped into so that a tool left on it will come tumbling down. I know it is tempting to leave the tool there for just a moment, but don't do it.

roof, and the top of the ladder was leaning against the flashing. There was a picket fence behind me and electrical wires entering the building to my right. The entrance wires were a good distance away, so I was in no immediate danger. As I worked on the ladder, a huge gust of wind blew around the building. I don't know where it came from; it hadn't been very windy when I went up the ladder.

The wind hit me and pushed the ladder and me sideways. The top of the ladder slid easily along the metal flashing, and I couldn't grab anything to stop me. I knew the ladder was going to go down, and I didn't have much time to make a decision. If I pushed off of the ladder, I would probably be impaled on the fence. If I rode the ladder down, it might hit the electrical wires and fry me. I waited until the last minute and jumped off the ladder.

I landed on the wet ground with a thud, but I missed the fence. The ladder hit the wires and sparks flew. Fortunately, I wasn't hurt and electricians were available to take care of the electrical problem. This was a case where I wasn't really negligent, but I could have been killed. If I had secured the top of the ladder, my accident wouldn't have happened.

Working Safely on a Ladder

1. Use a solid and level footing to set up the ladder.

2. Use a ladder in good condition; do not use one that needs repair.

3. Be sure step ladders are opened fully and locked.

4. When using an extension ladder, place it at least ¼ of its length away from the base of the building.

5. Tie an extension ladder to the building or other support to prevent it from falling or blowing down in high winds.

6. Extend a ladder at least 3 feet over the roof line.

7. Keep both hands free when climbing a ladder.

8. Do not carry tools in your pocket when climbing a ladder. (If you fall, the tools could cut into you and cause serious injury.)

9. Use the ladder the way it should be used. For example, do not allow two people on a ladder designed for use by one person.

10. Keep the ladder and all its steps clean—free of grease, oil, mud, etc.—in order to avoid a fall and possible injury.

FIGURE 15.5 Working safely on a ladder.

Screwdrivers and Chisels

Eye injuries and puncture wounds are common when working with screwdrivers and chisels. When the tools are used properly and safety glasses are worn, few accidents occur.

The key to avoiding injury with most hand tools is simply to use the right tool for the job. If you use a wrench as a hammer or a screwdriver as a chisel, you are asking for trouble. There are, of course, other types of tools and safety hazards found in the electrical trade. However, this list covers the ones that result in the most injuries. In all cases, observe proper safety procedures and utilize safety gear, such as eye and ear protection.

FIRE PREVENTION

Electricians need to be aware of the various fire hazards in their field. In addition to the possibility of exposed wiring coming in contact with flammable building materials, the use of welding equipment and propane torches presents additional fire hazards. Everyone on the job site should be familiar with the risk of fire and the location and use of fire extinguishers (Figure 15.6).

CO-WORKER SAFETY

Co-worker safety is the last segment of this chapter. I am including it because workers are frequently injured by the actions of co-workers. This section is meant to protect you from the mistakes of others and to make you aware of how your actions might affect your co-workers.

Most electricians find themselves working around other people. This is especially true on construction jobs. When working around other people, you must be aware of their actions, as well as your own. If you are walking out of a house to get something off the truck and a roll of roofing paper gets away from a roofer, you could get an instant headache.

If you don't pay attention to what is going on around you, it is possible to wind up in all sorts of trouble. Cranes lose their loads some times, and such a load landing on you is likely to be fatal. Equipment operators don't always see the electrician who is kneeling down. It's not hard to have a close encounter with heavy equipment.

To Prevent Fires

1. Always keep fire extinguishers handy, and be sure that the extinguisher is full and that you know how to use it quickly.

2. Be sure to disconnect and bleed all hoses and regulators used in welding, brazing, soldering, etc.

3. Store cylinders of acetylene, propane, oxygen, and similar substances in an upright position in a well-vented area.

4. Operate all air acetylene, welding, soldering, and related equipment according to the manufacturer's directions.

5. Do not use propane torches or other similar equipment near material that can easily catch fire.

6. Be careful at all times. Be prepared for the worst, and be ready to act.

FIGURE 15.6 Tips for fire prevention.

Always be aware of what is going on over your head. Avoid working under other people and hazardous overhead conditions. Let people know where you are, so you won't get stranded on a roof or in an attic when your ladder is moved or falls over.

You must also remember that your actions could harm co-workers. If you are on a roof and a hammer gets away from you, somebody could get hurt. Open communication between workers is one of the best ways to avoid injuries. If everyone knows where everyone else is working, injuries are less likely. Primarily, think and think some more. There is no substitute for common sense. Try to avoid working alone, and remain alert at all times.

16

FIRST AID

Everyone should invest some time in learning the basics of first aid. You never know when having skills in first aid treatment may save your life. Electricians live with what can be a dangerous profession. On-the-job injuries are not uncommon. Most injuries are fairly minor, but they often require treatment. Do you know the right way to get a sliver of copper out of your hand? If your helper suffers from an electrical shock when a drill cord goes bad, do you know what to do? Well, many electricians don't possess good first aid skills.

Before we get too far into this chapter, there are a few points I want to make. First of all, I'm not a medical doctor or any type of trained medical person. I've taken first aid classes, but I'm certainly not an authority on medical issues. The suggestions I will give you in this chapter are for informational purposes only. This book is not a substitute for first aid training offered by qualified professionals.

My intent here is to make you aware of some basic first aid procedures that can make life on the job much easier, but I want you to understand that I'm not advising you to use my advice to administer first aid. Hopefully, this chapter will show you the types of advantages you can gain from taking first aid classes. Before you attempt first aid on anyone, including yourself, you should attend a structured, approved first aid class. I'm going to give you information that is as accurate as I can make it, but don't assume that my words are enough. Take a little time to seek professional training in the art of first aid. You may never use what you learn. However, the one time it is needed, you will be glad you made the effort to learn what to do. With this said, let's jump right into some tips on first aid.

OPEN WOUNDS

Open wounds are a common problem for electricians. Many tools and materials used by electricians can create open wounds. What should you do if you or one of your workers is cut?

- *Stop the bleeding as soon as possible.*
- *Disinfect and protect the wound from contamination.*
- *You may have to take steps to avoid shock symptoms.*
- *Once the patient is stable, seek medical attention for severe cuts.*

When a bad cut is encountered, the victim may slip into shock. A loss of consciousness could result from a loss of blood. Death from extreme bleeding is also a risk. As a first-aid provider, you must act quickly to reduce the risk of serious complications.

Bleeding

To stop bleeding, direct pressure is normally a good tactic. This may be as crude as clamping your hand over the wound, but a cleaner compression is desirable. Ideally, a sterile material should be placed over the wound and secured, normally with tape (even if it's duct tape or the electrician's secret weapon, black tape). Whenever possible, wear rubber gloves to protect yourself from possible disease transfer if you are working on someone else. Thick gauze used as a pressure material can absorb blood and allow the clotting process to start.

Bad wounds may bleed right through the compress material. If this happens, don't remove the blood-soaked material. Add a new layer of material over it. Keep pressure on the wound. If you are not prepared with a first aid kit, you could substitute gauze and tape with strips cut from clothing that can be tied in place over the wound. A clean handkerchief or any piece of clean cloth would be ideal.

When you are dealing with a bleeding wound, it is usually best to elevate it. If you suspect a fractured or broken bone in the area of the wound, elevation may not be practical. When we talk about elevating a wound, it simply means to raise the wound above the level of the victim's heart. This helps slow the blood flow due to gravity.

Severe Bleeding

Severe bleeding might not stop even after a compression bandage is applied and the wound is elevated. When this is the case, you must resort to putting pressure on the main artery that is producing the blood. Constricting an artery is not an alternative for the steps that we have discussed previously.

Putting pressure on an artery is serious business. First, you must be able to locate the artery, and you should not keep the artery constricted any longer than necessary. You may have to apply pressure for a period of time, release it, and then apply it again. It's important that you do not restrict the flow of blood in arteries for long periods of time. I hesitate to go into too much detail on this process, as I feel it is a method that you should be taught in a controlled classroom situation. However, I will hit the high spots. But, remember, these words are not a substitute for professional training from qualified instructors.

Open arm wounds are controlled with the brachial artery. The location of this artery is in the area between the biceps and triceps, on the inside of the arm. It's about halfway between the armpit and the elbow. Pressure is created with the flat parts of your fingertips. Basically, you are holding the victim's wrist with one hand and closing off the artery with your other hand. Pressure exerted by your fingers pushes the artery against the arm bone and restricts blood flow. Again, don't attempt this type of first aid until you have been trained properly in the execution of the procedure.

Severe leg wounds may require the constriction of the femoral artery. This artery is located in the pelvic region. Normally, bleeding victims are placed on their backs for this procedure. The heel of a hand is placed on the artery to restrict blood flow. In some cases, fingertips are used to apply pressure. I'm uncomfortable with going into great detail on these procedures, because I don't want you to rely solely on what I'm telling you. It's enough that you understand the fact that knowing when and where to apply pressure to arteries can save lives and that you should seek professional training in these techniques.

Tourniquets

Tourniquets get a lot of attention in movies, but they can do as much harm as good if not used properly. A tourniquet should only be used in a life-threatening situation. When a tourniquet is applied, there is a risk of losing the limb to which the restriction is applied. This is obviously a serious decision and one that must be made only when all other means of stopping blood loss have been exhausted.

Unfortunately, electricians might run into a situation where a tourniquet is the only answer. For example, if a worker allowed a power saw to get out of control, a hand might be severed or some other type of life-threatening injury could occur. This would be cause for the use of a tourniquet. Let me give you a basic overview of what's involved when a tourniquet is used.

Tourniquets should be at least two inches wide. A tourniquet should be placed at a point that is above a wound, between the bleeding and the victim's heart. However, the binding should not encroach directly on the wound area. Tourniquets can be fashioned out of many materials. If you are using strips of cloth, wrap the cloth around the limb that is wounded and tie a knot in the material. Use a stick, screwdriver, or whatever else you can lay your hands on to tighten the binding.

Once you have made a commitment to apply a tourniquet, the wrapping should only be removed by a physician. It's a good idea to note the time that a tourniquet is applied, as this will help doctors later in assessing their options. As an extension of the tourniquet treatment, you will most likely have to treat the patient for shock.

Infection

Infection is always a concern with open wounds. When a wound is serious enough to require a compression bandage, don't attempt to clean the cut. Keep pressure on the wound to stop bleeding. In cases of severe wounds, be on the lookout for shock symptoms and be prepared to treat them. Your primary concern with a serious open wound is to stop the bleeding and gain professional medical help as soon as possible.

Lesser cuts, which are more common than deep ones, should be cleaned. Regular soap and water can be used to clean a wound before applying a bandage. Remember, we are talking about minor cuts and scrapes at this point. Flush the wound generously with clean water. A piece of sterile gauze can be used to pat the wound dry. Then a clean, dry bandage can be applied to protect the wound while in transport to a medical facility.

SPLINTERS

Splinters and foreign objects often invade the skin of electricians. Getting these items out cleanly is best done by a doctor, but there are some on-the-job methods that you might want to try. A magnifying glass and a pair of tweezers work well together when removing embedded objects, such as splinters and slivers of copper or aluminum wire. Ideally, tweezers should be sterilized either over an open flame, such as the flame of your torch, or in boiling water.

Splinters and slivers that are submerged beneath the skin can often be lifted out with the tip of a sterilized needle. The use of a needle in conjunction with a pair of tweezers is very effective in the removal of most simple splinters. If you are dealing with something that has gone extremely deep into tissue, it is best to leave the object alone until a doctor can remove it.

Quick Review of First Aid Techniques

- ✔ Use direct pressure to stop bleeding.
- ✔ Wear rubber gloves to prevent direct contact with a victim's blood.
- ✔ When feasible, elevate the part of the body that is bleeding.
- ✔ Extremely serious bleeding can require you to put pressure on the artery supplying the blood to the wound area.
- ✔ Tourniquets can do more harm than good.
- ✔ Tourniquets should be at least two inches wide.
- ✔ Tourniquets should be placed above the bleeding wound, between the bleeding and the victim's heart.
- ✔ Tourniquets should not be applied directly on the wound area.
- ✔ Tourniquets should be removed only by trained medical professionals.
- ✔ If you apply a tourniquet, note the time that you apply the tourniquet.
- ✔ When a bleeding wound requires a compression bandage, don't attempt to clean the wound. Simply apply compression quickly.
- ✔ Watch victims with serious bleeding for symptoms of shock.
- ✔ Minor wounds should be cleaned before being bandaged.

EYE INJURIES

Eye injuries are very common on construction and remodeling jobs. Most of these injuries could be avoided with proper eye protection, but far too many workers don't wear safety glasses and goggles. This sets the stage for eye irritations and injuries.

Before you attempt to help someone who is suffering from an eye injury, you should wash your hands thoroughly. I know this is not always possible on construction sites, but cleaning your hands is advantageous. In the meantime, keep the victim from rubbing the injured eye. Rubbing can make matters much worse.

Never attempt to remove a foreign object from someone's eye with the use of a rigid device, such as a toothpick. Cotton swabs that have been wetted can serve well as a magnet to remove some types of invasion objects. If the person you are helping has something embedded in an eye, get the person to a doctor as soon as possible. Don't attempt to remove the object yourself.

When you are investigating the cause of an eye injury, you should pull down the lower lid of the eye to determine if you can see the object causing trouble. A floating object, such as a piece of sawdust trapped between an eye and an eyelid can be removed with a tissue, a damp cotton swab, or even a clean handkerchief. Don't allow dry cotton material to come into contact with an eye.

If looking under the lower lid doesn't pinpoint the source of discomfort, check under the lower lid. Clean water can be used to flush out many eye contaminants without much risk of damage to the eye. Objects that cannot be removed easily should be left alone until a physician can take over.

I am reminded of the time I got a splinter of steel in the pupil of my eye. I was making a three inch hole up through the steel decking below a large motor control center. I was using a hole saw above my head. I was wearing safety glasses, but I had my head down to shield my face from metal filings produced by the hole saw. A splinter of steel shot down and ricocheted off from the inside of the lens of my safety glasses directly into the center of my left eye. The pain was excruciating.

An old electrician grabbed me and pulled me into the light. He then took out a soft paper match and rubbed the torn end back and forth over the steel splinter until he retrieved it from my eye. Then he took me to the office trailer where I was immediately transported to the emergency room at the local hospital. He acted quickly and calmly; and he kept me calm. Now, that's what I call first aid.

Eye injuries need immediate and proper attention. When an eye injury occurs it's important to remember the following:

- *Wash your hands, if possible, before treating eye injuries.*
- *Don't rub an eye wound.*
- *Don't attempt to remove embedded items from an eye.*
- *Clean water can be used to flush out some eye irritants.*

SCALP INJURIES

Scalp injuries can be misleading. What looks like a serious wound can be a fairly minor cut. On the other hand, what appears to be only a cut can involve a fractured skull. If you or someone around you sustains a scalp injury, such as having a hammer fall on your head from an overhead worker, take it seriously. Don't attempt to clean the wound. Expect profuse bleeding.

If you don't suspect a skull fracture, raise the victim's head and shoulders to reduce bleeding. Try not to bend the neck. Put a sterile bandage over the wound, but don't apply excessive pressure. If there is a bone fracture, pressure could worsen the situation. Secure the bandage with gauze or some other material that you can wrap around it. Seek medical attention immediately.

FACIAL INJURIES

Facial injuries can occur on electrical jobs. I've seen helpers let their right-angle drills get away from them resulting in hard knocks to the face. On one occasion, I remember a tooth being lost. Split lips and bitten tongues are common when a drill goes on a rampage.

Extremely bad facial injuries can cause a blockage of the victim's air passages. This, of course, is a very serious condition. It's critical that air passages are open at all times. If the person's mouth contains broken teeth or dentures, remove them. Be careful not to jar the individual's spine if you have reason to believe there may be injury to the back or neck.

Conscious victims should be positioned, when possible, so that secretions from the mouth and nose will drain out. Shock is a potential concern in severe facial injuries. For most on-the-job injuries, victims should be treated for comfort and sent for medical attention.

NOSEBLEEDS

Nosebleeds are not usually difficult to treat. Typically, pressure applied to the side of the nose where bleeding is occurring will stop the flow of blood. Applying cold compresses can also help. If external pressure does not stop the bleeding, use a small, clean pad of gauze to create a dam on the inside of the nose. Then, apply pressure on the outside of the nose. This will almost always work. If it doesn't, get to a doctor.

BACK INJURIES

There is really only one thing that you need to know about back injuries. Don't move the injured party. Call for professional help and see that the victim remains still until help arrives. Moving someone who has suffered a back injury can be very risky. Don't do it unless there is a life-threatening cause for your action, such as a person trapped in a fire or some other type of deadly situation.

LEGS AND FEET

Legs and feet sometimes become injured on job sites. The worst case of this type that I can remember was when a plumber knocked a pot of molten lead over on his foot. It sends shivers up my spine just to recall that incident. Anyway, when someone suffers a minor foot or leg injury, you should clean and cover the wound. Bandages should be supportive without being constrictive. The appendage should be elevated above the victim's heart level when possible. Prohibit the person from walking. Remove boots and socks so that you can keep an eye on the person's toes. If the toes begin to swell or turn blue, loosen the supportive bandages.

Blisters

Blisters may not seem like much of an emergency, but they can sure take the steam out of a helper or electrician. In most cases, blisters can be covered with a heavy gauze pad to reduce pain. It is generally recommended to leave blisters unbroken. When a blister breaks, the area should be cleaned and treated as an open wound. Some blisters tend to be more serious than others. For example, blisters in the palm of a hand or on the sole of a foot should be examined by a doctor. The main causes of blisters for an electrician are pulling cable and threading conduit together with bare hands. A simple preventive measure for this is to wear gloves.

HAND INJURIES

Hand injuries are common in the electrical trade. Little cuts are the most frequent complaint. Serious hand injuries should be elevated. This tends to reduce swelling. You should not try to clean really bad hand injuries. Use a pressure bandage to control bleeding. If the cut is on the palm of a hand, a roll of gauze can be squeezed by the victim to slow the flow of blood. Pressure should stop the bleeding, but if it doesn't, seek medical assistance.

As with all injuries, use common sense on whether or not professional attention is needed after first aid is applied. The biggest cause of cuts to the hands and wrists is by raking them across a sharp object. This usually occurs when a drill bit finally cuts through the material being drilled and the hand is plunged forward over the sharp edge of a cabinet. Another common hazard is a partially driven nail or screw that is unseen. Pulling cable around these hazards will make a nasty cut if they make contact with your hands when you give a big pull on the cable.

SHOCK

Shock is a condition that can be life threatening even when the injury is not otherwise fatal. We are talking about traumatic shock, not electrical shock. Many factors can lead to a person going into shock. A serious injury is a common cause, but many other causes exist. There are certain signs of shock that you can look for.

If a person's skin turns pale or blue and is cold to the touch, it's a likely sign of shock. Skin that becomes moist and clammy can indicate the presence of shock. General weakness is also a sign of shock. When a person is going into shock, the pulse rate is likely to exceed 100 beats per minute. Breathing is usually increased, but it may be shallow, deep, or irregular. Chest injuries usually result in shallow breathing. Victims who have lost blood may be thrashing about as they enter into shock. Vomiting and nausea can also signal shock.

As a person slips into deeper shock, the individual may become unresponsive. Look at the eyes; they may be widely dilated. Blood pressure can drop, and in time, the victim will lose consciousness. Body temperature will fall, and death will be likely if treatment is not rendered.

There are three main goals when treating someone for shock. First, get the person's blood circulating well. Second, make sure an adequate supply of oxygen is available to the individual. Third, maintain the person's body temperature.

When you have to treat a person for shock, you should keep the victim lying down. Cover the individual so that the loss of body heat will be minimal. Get medical help as soon as possible. The reason for keeping a person lying down is to slow the circulation of blood. Remember, if you suspect back or neck injuries, don't move the person.

People who are unconscious should be placed on one side so that fluids will run out of the mouth and nose. It's also important to make sure that air passages are open. A person with a head injury may be laid out flat or propped up, but the head should not be lower than the rest of the body. It is sometimes advantageous to elevate a person's feet when they are in shock. However, if there is any difficulty in breathing, or if pain increases when the feet are raised, lower them.

Body temperature is a big concern with shock patients. You want to avoid chilling. However, don't attempt to add additional heat to the surface of the person's body with artificial means. This can be damaging. Use only blankets, clothes, and other similar items to regain and maintain body temperature.

Avoid the temptation to offer the victim fluids, unless medical care is not going to be available for a long time. Avoid fluids completely if the person is unconscious or is subject to vomiting. Under most job-site conditions, fluids should not be administered.

Checklist of shock symptoms:

✔ Skin that is pale, blue, or cold to the touch

✔ Skin that is moist and clammy

✔ General weakness

✔ Pulse rate in excess of 100 beats per minute

✔ Increased breathing

✔ Shallow breathing

✔ Thrashing

✔ Vomiting and nausea

✔ Unresponsive action

✔ Widely dilated eyes

✔ Drop in blood pressure

BURNS

Burns are not very common among electricians, but they can occur in the workplace. There are three types of burns that you may have to deal with. First-degree burns are the least serious. These burns typically come from overexposure to the sun, quick contact with a hot object, or contact with scalding water from working with a boiler or water heater.

Second-degree burns are more serious. They can come from a deep sunburn or from contact with hot liquids and flames. A person who is affected by a second-degree burn may have a red or mottled appearance, blisters, and a wet appearance of the skin within the burn area. This wet look is due to a loss of plasma through the damaged layers of skin.

Third-degree burns are the most serious. They can be caused by contact with open flames, hot objects, or immersion in very hot water. Electrical injuries can also result in third-degree burns. This type of burn can look similar to a second-degree burn, but the difference will be the loss of all layers of skin.

Treatment

Treatment for most job-related burns can be administered on the job site and will not require hospitalization. First-degree burns should be washed with or submerged in cold water. A dry dressing can be applied if necessary. These burns are not too serious. Eliminating pain is the primary goal with first-degree burns.

Second-degree burns should be immersed in cold (but not ice) water. The soaking should continue for at least one hour and up to two hours. After soaking, the wound should be layered with clean cloths that have been dipped in ice water and wrung out. Then the wound should be dried by blotting, not rubbing. Dry, sterile gauze should then be applied. Don't break open any blisters. It is also not advisable to use ointments and sprays on severe burns. Burned arms and legs should be elevated, and medical attention should be the first priority.

Bad burns, the third-degree type, need quick medical attention. First, don't remove a burn victim's clothing; skin might come off with it. A thick, sterile dressing can be applied to the burn area. Personally, I would avoid this if possible. A dressing might stick to the mutilated skin and cause additional skin loss when the dressing is removed. When hands are burned, keep them elevated above the victim's heart. The same goes for feet and legs. You should not soak a third-degree burn in cold water; it could induce more shock symptoms. Don't use ointments, sprays, or other types of treatments. Get the burn victim to competent medical care as soon as possible. Most burns for electricians occur when applying heat-shrink or bending PVC conduit. A simple preventive measure is to wear gloves and to have cold water handy.

HEAT-RELATED PROBLEMS

Heat-related problems can include heat stroke and heat exhaustion. Cramps are also possible when working in hot weather. There are people who don't consider heat stroke to be serious. They are wrong. Heat stroke can be life threatening. People affected by heat stroke can develop body temperatures in excess of 106°F. Their skin is likely to be hot, red, and dry. You might think sweating would take place, but it doesn't. Pulse is rapid and strong, and victims can sink into an unconscious state.

If you are dealing with heat stroke, you need to lower the person's body temperature quickly. There is a risk, however, of cooling the body too quickly once the victim's temperature is below 102°F. You can lower body temperature with rubbing alcohol, cold packs, cold water on clothes, or in a bathtub of cold water. Avoid the use of ice in the cooling process. Fans and air-conditioned space can be used to achieve your cooling goals. Get the body temperature down to at least 102° and then go for medical help. A neat way to stay cool while working in excessive temperatures is to set up a box fan a few feet away from you. They are inexpensive and work great. This isn't always practical but I have used this method for years and it makes working in extremely hot weather like a vacation.

Cramps

Cramps are not uncommon among workers during hot spells. A simple massage can be all it takes to cure this problem. Saltwater solutions are another way to control cramps. Mix one teaspoonful of salt per glass of water and have the victim drink half a glass about every 15 minutes.

Exhaustion

Heat exhaustion is more common than heat stroke. A person affected by heat exhaustion is likely to maintain a fairly normal body temperature, but the person's skin may be pale and clammy. Sweating may be very noticeable, and the individual will probably complain of being tired and weak. Headaches, cramps, and nausea may accompany the symptoms. In some cases, fainting might occur.

The salt-water treatment described for cramps will normally work with heat exhaustion. Victims should lie down and elevate their feet about a foot off the floor or bed. Clothing should be loosened, and cool, wet cloths can be used to add comfort. If vomiting occurs, get the person to a hospital for intravenous fluids.

We could continue talking about first aid for a much longer time. However, the help I can give you here for medical procedures is limited. You owe it to yourself, your family, and the people you work with to learn first aid techniques. This can best be done by attending formal classes in your area. Most towns and cities offer first-aid classes on a regular basis. I strongly suggest that you enroll in one. Until you have some hands-on experience in a classroom and gain the depth of knowledge needed, you are not prepared for emergencies. Don't get caught short. Prepare now for the emergency that might never happen.

GLOSSARY

accent lighting Lighting that is aimed at a specific location, such as a decoration or architectural element.

alternating current (AC) Electrical current that regularly reverses polarity from positive to negative, usually at 60 cycles per second.

ambient lighting Indirect or background lighting.

ampere (amp) The rate at which electrical power flows to a fixture, tool, or appliance.

ampacity Current-carrying capacity of an electrical wire, measured in amps.

armored cable Wires that are grouped together and protected by a flexible metal covering.

backfeed A dangerous condition where electrical current is fed back into a utility system while a standby generator is running.

ballast A transformer that steps up the voltage in a fluorescent lamp. Ballasts also limit current in a fluorescent lamp.

bar hanger A device that is installed between framing joists or rafters to support a fixture box, such as one for a ceiling fan.

bare wire A wire that has had the insulation stripped from it or a wire that has been manufactured without any insulation.

base pin Contact on the end of a fluorescent tube.

bell wire Thin wire, normally 18 gauge, which is used for doorbells.

bipin Fluorescent tube that has two base pins on either end.

bimetal Two metals that heat and cool at different rates to open or close a circuit automatically.

bonding The act of connecting metal parts to make an electrically conductive path, usually to ground.

bonding jumper A connection that maintains constant conductivity between metal parts required to be electrically connected.

box A device used to contain wiring connections and devices.

box extension A device that is attached to an electrical box to increase the capacity of the box or to make it flush with the finished surface.

branch circuit Wiring that runs from a final fuse or circuit breaker to receptacles or other loads.

brownout A reduction in electric power.

bus bar A long terminal inside of a service panel. Circuit breakers and fuses connect to bus bars to distribute power.

BX cable The same as armored cable.

cable Wires that are grouped together and protected by a covering or sheath.

choke ballast A ballast that lacks a transformer and which is used only in small fluorescent fixtures.

circuit A continuous loop of electrical current that flows along wires or cables.

circuit breaker A safety device that interrupts an electrical circuit if the circuit becomes overloaded or shorted.

circuit capacity The maximum current a circuit can safely handle.

coaxial cable A primary conductor wire that is enclosed in concentric plastic foam insulation. A secondary conductor is created by the braided wire that covers the primary conductor to shield against interference.

common application language (CAL) A language that allows several household systems to be combined under a single system of control.

common terminal On a three-way switch, the darkest terminal to which the wire supplying power is connected.

common wire In a three-way switch setup, the wire that brings power to the switch. Also the load wire that takes power to the fixture.

communicating thermostat A thermostat that is interactive and that can be remotely controlled.

conductor Material that allows electrical current to flow through it.

conductor cable NM sheathing encasing multiple conductors.

conduit A protective tube-like device in which wiring can be installed. Some conduit is made of metal and other conduit is made of plastic.

conduit nipple A short section of conduit used to connect an interior junction box to an outdoor receptacle box, an LB connector, or two enclosures, usually 24 inches or less.

connecting block A central distribution junction for telephone circuits.

contact A point where two electrical conductors come together.

continuity Uninterrupted electrical pathway through a circuit or fixture.

continuity tester A tool that can indicate if a circuit is capable of carrying electricity.

continuous load A load where maximum current continues for three hours or more.

controller A device that controls the electric power delivered to another device to which it is directly or remotely connected.

cord Multiple insulated strands of wires that are encased in sheathing, usually flexible.

crimping ferrule A compression sleeve that is used to connect bare grounding wires.

current Movement of electricity along a conductor.

cut-in box An electrical box that can be easily installed in existing construction.

delayed-start tube A type of fluorescent tube that takes a few seconds to warm up.

design protocol A control standard that is used for home automation devices.

detector A device that senses changes in ambient conditions due to smoke, gas, temperature, motion, flame, and similar elements.

device Usually an electrical receptacle or switch.

digital satellite system (DSS) A system that distributes video signals via satellite to an antenna dish receiver.

digital versatile disk (DVD) Formerly digital videodisc; digital movie disc format having multiple layers.

dimmer A switch that allows the intensity of a light to be varied by operating the switch.

direct current (DC) Electrical current that flows only in one direction.

double-pole switch An electrical switch that has two blades and contacts to alternately open or close power.

double-pole double-throw switch An electrical switch that has two blades and contacts to alternately open or close two sources of power.

dry niche A fixture that is not waterproof and that houses a pool light.

duplex receptacle An electrical receptacle, or outlet, that accepts up to two electrical plug ends.

electrical metallic tubing (EMT) A thin, unthreaded metal conduit that can be used for applications where not subject to severe physical damage.

electrical nonmetallic tubing PVC conduit that must be concealed behind finished surfaces.

electrons Invisible particles of charged matter moving at the speed of light through an electrical circuit.

end-line wiring (switch-loop wiring) A method of wiring a switch in which power runs to the fixture box.

end-of-run The last receptacle on a circuit.

energy efficiency rating (ERR) A measure of relative energy consumption.

equipment grounding conductor A conductor that connects non-current-carrying metal parts of equipment to a grounding conductor and/or grounding electrode conductor.

escutcheon plate A protective plate.

expansion loop Slack provided in a cable to allow for expansion and contraction.

extension ring A device that can be installed on an electrical box to expand the capacity of the box.

feedhorn A device at the focal point of a satellite dish that receives signals reflected from the dish.

feed wire A conductor that carries 120 volt current uninterrupted from a service panel.

fish tape A long strip of spring steel used for fishing cables and for pulling wires and cables through conduit.

fixed-temperature detector A heat detector that uses the low melting point of solder or metals that expand when exposed to heat or fire.

fixture A light or fan that is permanently attached (hard wired) to an electrical system.

flame detector A detector that senses flames.

flexible metal conduit Tubing that can be bent easily by hand.

fluorescent tube A light source that uses an ionization process to produce ultraviolet radiation that becomes visible light when it hits the coated inner surface of the tube.

four-way switch A switch used when a light is controlled by three or more switches.

fuse A safety device that interrupts electrical circuits during an overload or short.

ganging Putting multiple electrical components in a single box.

greenfield The same as armored cable but without any factory-installed conductors.

grounded wire A wire that returns current at zero voltage to the source of electrical power. This type of wire is usually covered with white insulation. Gray insulation and insulation with three white traces are also acceptable.

grounding wire A wire used in an electrical circuit to conduct current to the earth in the event of a short circuit. This wire is usually a bare copper wire. Green insulation and green insulation with three yellow traces are also acceptable.

ground-fault-circuit interrupter (GFCI or GFI) A receptacle with a built-in safety feature that shuts a circuit off when there is a breech that may produce a risk of shock. GFIs are designed to shut off at a ground fault of 4 to 6 milliamps maximum.

grounding rod A metal rod that acts as a conductor and that is buried in the earth to maintain ground potential on other conductors that are attached to it.

hanger bracket An adjustable bracket that a ceiling box or fixture is hung from.

hardwired A method of wiring a fixture directly to an electrical system, rather than using a corded plug to plug the fixture into an outlet.

heavy-duty circuit A circuit that serves a single appliance.

hertz A unit of frequency measuring one cycle per second.

hickey A threaded fitting used to connect a light fixture to a ceiling box. A rigid conduit hand bender is also called a hickey.

home automation A remote-controlled system designed to control a variety of devices in a home.

home networking Complete home connectivity.

hot bus A metal bar in an electrical panel that serves as a common connection between circuit breakers and hot line conductors.

hot wire A wire that carries voltage. This wire will normally have red or black insulation; in addition, any solid color other than white, gray, or green can serve as a hot wire.

incandescent bulb A light source with an electrically charged metal filament that burns at white heat.

impedance Opposition to current flow in an AC circuit.

instant-start starter A device used to make fluorescent lamps produce light faster.

insulation A nonconductive covering that protects wires and other carriers of electricity.

insulation-contact A recessed light fixture housing approved for direct contact with insulation.

insulation displacement connector (IDC) A telephone circuit junction-wiring block having a gas-tight seal that prevents bimetal corrosion.

insulator A material that resists the flow of electrical current.

IR Infrared.

integral transformer Low-voltage transformer that is built into a device.

interactive television (ITV) Television that permits user interaction.

intermediate metallic conduit (IMC) Threaded rigid conduit that has thin walls.

ionization detector A sensor that ionizes the air between electrodes, causing a current to flow. Smoke particles interfering with this flow set off an alarm.

isolated-ground receptacle An orange receptacle that is wired to an independent grounding system that protects sensitive electronic equipment from electrical noise. These receptacles are common in medical facilities and for computers.

jumper wire A short piece of wire that is used to complete a circuit between contacts.

junction The point where wire splices occur.

junction box *See box.*

kilowatt (kW) 1,000 watts.

kilowatt-hour (kWh) The amount of energy expended in one hour by one kilowatt of electricity.

knockout A round slug or a tab that can be punched out of a box or panel to allow the installation of a cable.

LB fitting A pulling elbow made for outdoor use.

lamp A device that generates light.

laser disk (LD) High-resolution disc on which programs can be recorded for playback on a television set.

lead A short wire that is connected to a fixture.

lightning rod A grounded metal rod placed on a structure to prevent damage by conducting lightning to the ground.

line cord A flat, four-conductor cord that is used to connect a telephone or other accessory to a phone jack.

load A device or equipment to which power is delivered.

local area network (LAN) A system that links electronic equipment to a shared network.

low-voltage Voltage stepped down from 120 volts to 30 volts, or less.

low-voltage transformer An electrical device with multiple coupled windings that step standard 120 volt power down to 30 volts, or less.

lumen The flow rate of light per unit of time, defining quantity of visible light.

luminaire A light fixture complete with lamps and ballast (where applicable).

luminosity The relative brightness of a light source.

magnetic flux The number of magnetic lines of force passing through a bounded area in a magnetic field.

main breaker A circuit breaker through which utility power enters the main panel and connects to the hot bus bar.

MC cable Armored cable with a ground wire in addition to at least two insulated wires.

meter An electrical tool used to measure the amount of electrical power being used.

middle-of-run A receptacle that is located between a service panel and another receptacle.

motion-sensor detector A passive infrared detector that senses movement.

multitester A tool that measures voltage of various levels, tests for continuity, and performs other tests.

nailing spur A nailing bracket attached to an electrical box.

network Connection between computers, peripheral equipment, and communications devices that permits the sharing of files, programs, and equipment.

network interface device A telephone utility device that connects house wiring to a telephone network.

neutral bus bar A bus bar that connects a utility neutral wire to a house neutral.

neutral wire A wire that returns current at zero voltage to the source of electrical power. This type of wire is usually covered with white or gray insulation or black with three white traces.

noninsulation-contact fixture A recessed light fixture housing that is not approved for direct contact with insulation.

nonmetallic cable (NM) Multiple wires contained in a sheath that is not made of metal.

non-polarized Two positions or poles that are not exclusively positive or negative.

ohm A unit of electrical resistance in a conductor.

Ohm's Law Current in a circuit is directly proportional to voltage and inversely proportional to resistance.

occupancy detector A detector that reacts to a variable such as heat and motion to detect the presence of people.

overcurrent A current that exceeds the rated current of equipment or the ampacity of a conductor.

overload Excessive demand for power on a circuit.

outlet Any opening in an electrical circuit for the purpose of power consumption such as lights, fans or receptacles.

overload A demand for more current than a circuit or device can accommodate.

parabolic aluminized reflector (PAR) A bulb that has an internal reflector of aluminum.

passive infrared (PIR) detector A device that senses body heat.

photocell A device that has an electrical output that varies in response to invisible light.

photoelectric detector A sensor that responds to changes in light levels.

pigtail A short wire that is used to connect multiple wires to a single terminal.

plug A fitting that has metal prongs for insertion into a fixed receptacle that is used to connect a cord to a power supply.

plug configuration The number and pattern of prongs on a plug.

point-of-use protection Protection that is created at the location of a device. A surge-suppressor outlet bar is a good example of point-of-use protection.

polarized receptacle A receptacle designed to keep hot current flowing along hot wires and neutral current flowing down neutral wires.

power The result of hot current flowing for a period of time.

power horn A security alarm.

preheat starter A device that preheats the electrodes of a fluorescent lamp.

programmable switch A switch that can be programmed to operate at designated times.

pulse A brief and sudden change in a normally constant current.

quad-shielded cable Coaxial cable that has two layers of foil shielding, each of which is covered by a layer of braided shielding.

raceway Surface-mounted channels made of plastic or metal that wire can be run through to create a circuit.

rapid-start starter A device that uses low-voltage winding to preheat fluorescent lamp electrodes.

rapid-start tubes Fluorescent tubes that light up almost instantly.

receptacle An electrical device that provides plug in access to electrical power.

recessed can light A light fixture that contains its own electrical box that is designed to be installed in a ceiling so that the housing trim is flush with the ceiling.

reflector lamp An incandescent lamp that has a reflector built into the bulb.

register Transmitting and recording information from a controlled device to a central controller.

remodel box An electrical box that clamps to a wall surface, rather than being attached to bare framing members.

resistance Opposition to current flow in a conductor, device, or load measured in ohms.

rigid conduit Metal conduit tubing that requires a bender to create bends and offsets.

Romex A brand name of plastic-sheathed electrical cable that is normally used for interior wiring requirements.

run wattage The power required to keep an appliance running.

screw terminal A screw and contact point on an electrical device, such as a switch or outlet, where wires are attached to the device.

secondary winding Transformer coils wound on the output side of a transformer.

service drop An overhead utility line that brings electric service to a home.

service entrance The location where power from a utility company enters a building.

service panel A metal box that receives the service cable from an electrical meter. Service panels house connection bars where electrical current can be split into individual circuits. Circuit breakers or fuses are also housed in service panels.

short circuit The result of an improper contact between two current-carrying wires or between a current-carrying wire and a grounding conductor.

single-pole switch An electrical switch that has one fixed contact and one movable contact.

small-appliance circuit A circuit that usually has two or three 20 amp receptacles.

smoke detector An ionization or photoelectric device that sounds an alarm when the device senses products of combustion.

splice A method used to connect two or more wires with a wire nut. Splices may also be made with mechanical lugs, compression connectors or by thermal weld.

starter switch A switch that works with a ballast to start a fluorescent tube when sufficient power is present.

start-of-run The first switch or outlet on a circuit.

stranded wire Wires that are spun together to form a single conductor.

stripping The removal of insulation from wire or sheathing from cable.

subpanel A subsidiary service panel that contains circuit breakers or fuses and supplies a number of branch circuits.

switch A device that controls electrical current passing through hot wires.

system ground A method by which an entire electrical system is grounded.

task lighting Lighting that is directional for point-of-use operation.

three-way switch A switch used when two switches control a light.

through-switch wiring (in-line wiring) Wiring a switch so that power runs to the switch box.

time-delay fuse A fuse that does not break the circuit during the momentary overload that can happen when an electric motor starts up. However, the fuse will blow if the overload continues.

transfer switch A device that switches a load from its main source to a standby power source. The switching action may be automatically or manually operated.

transformer A piece of electrical equipment that changes voltage in a circuit. If the voltage is increased it is called a step-up transformer. If the voltage is decreased it is called a step-down transformer.

transponder A control device that receives a remote signal and then sends out its own signal to another device.

travelers Two conductors that run between switches in a three-way installation. They are connected to the two like colored terminals on the switches.

twistlock A receptacle or plug that can be locked in place to prevent accidental removal.

type S fuse A plug fuse that can only fit into a fuse holder or adapter having the same amperage rating.

UL An abbreviation for Underwriters Laboratories.

ultrasonic sensor A device that emits high frequency sound and then listens for changes in the echo.

ultraviolet (UV) light Range of invisible light just beyond violet in the spectrum.

Underwriter's knot A knot that ensures electrical wiring connections in a fixture will withstand the strain on a cord that is jerked.

voltage (volts) A measurement of electricity in terms of pressure.

voltage detector A tool that senses electrical pressure.

voltage tester A tool that senses the presence of electrical pressure when its probes touch bare wire ends. Some voltage testers are capable of telling how many volts are present.

watertight A term given to devices that are designed to be temporarily immersed in water.

wattage (watt) A measurement of electrical power in terms of total energy consumed. A product of voltage and amperage. Watts = volts x amps.

weatherproof A means of construction to allow a device to function without interference from weather.

wire connector A plastic device with a metal spring on the interior of the connector that is used to connect multiple wires safely.

zip cord A two-wire cord that is made to split down the middle when pulled apart.

APPENDIX: ELECTRICAL SYMBOLS AND DATA

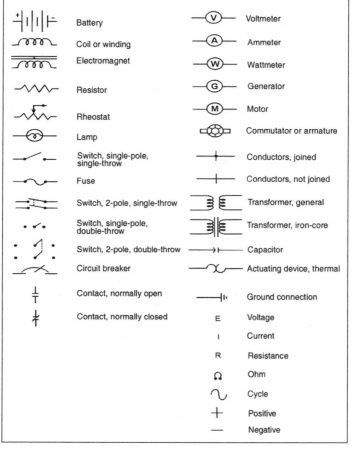

Symbol	Name	Symbol	Name
	Battery	Voltmeter	Voltmeter
	Coil or winding	Ammeter	Ammeter
	Electromagnet	Wattmeter	Wattmeter
	Resistor	Generator	Generator
	Rheostat	Motor	Motor
	Lamp	Commutator or armature	Commutator or armature
	Switch, single-pole, single-throw	Conductors, joined	Conductors, joined
	Fuse	Conductors, not joined	Conductors, not joined
	Switch, 2-pole, single-throw	Transformer, general	Transformer, general
	Switch, single-pole, double-throw	Transformer, iron-core	Transformer, iron-core
	Switch, 2-pole, double-throw	Capacitor	Capacitor
	Circuit breaker	Actuating device, thermal	Actuating device, thermal
	Contact, normally open	Ground connection	Ground connection
	Contact, normally closed	E	Voltage
		I	Current
		R	Resistance
		Ω	Ohm
			Cycle
		+	Positive
		—	Negative

FIGURE A.1   Basic circuiting symbols.

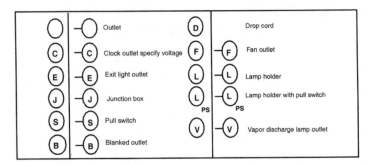

FIGURE A.2 General outlet symbols.

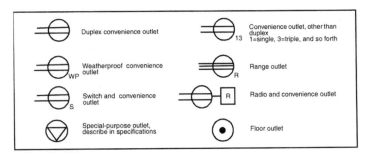

FIGURE A.3 Convenience outlet symbols.

S	Single-pole switch	S_4	Four-way switch
S_3	Three-way switch	S_E	Electrolier switch
S_D	Automatic-door switch	S_P	Pilot lamp and switch
S_K	Key-operated switch	S_{WCB}	Weatherproof circuit breaker
S_{CB}	Circuit breaker	S_{RC}	Remote-control switch
S_{MC}	Momentary contact switch	S_F	Fused switch
S_{WP}	Weatherproof switch	S_{WF}	Weatherproof fused switch
S_2	Double-pole switch		

FIGURE A.4 Switch outlet symbols.

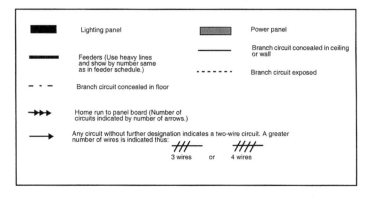

FIGURE A.5   Panels and circuits.

Push button	Buzzer	Bell
Electric door opener	Fire-alarm station	Fire-alarm bell
Controller	Horn	Nurse's signal plug
Isolating switch	Radio outlet	Bell-ringing transformer
Annunciator		

FIGURE A.6   Miscellaneous symbols.

Type	Wiring Diagram	Voltage	Use
A Single-phase, two-wire 1Ø 2 W	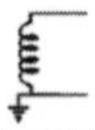	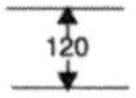 120	Lighting and small, single-phase motors, small loads
B Single-phase, three-wire 1Ø 3 W		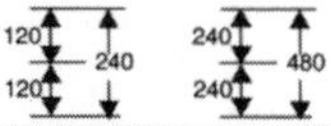 120 / 240 / 120 240 / 480 / 240	Local power to small buildings
C Three-phase, four-wire 3Ø 4 W	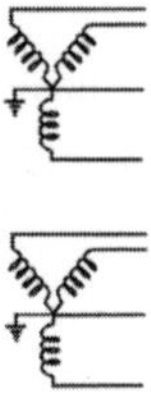	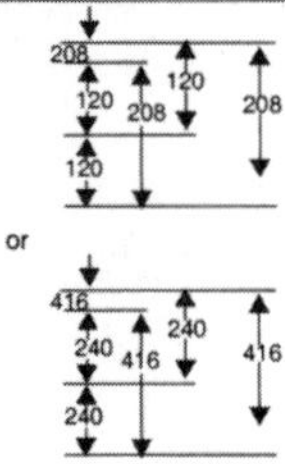 208 / 120 / 208 / 120 / 208 208 or 416 / 240 / 416 / 240 / 416 416	Most common system for military secondary distribution
D Three-phase, three-wire 3Ø 3 W		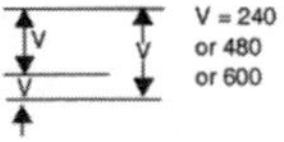 V V = 240 or 480 or 600	Large motor loads, small lighting loads
E Three-phase, four-wire 3Ø 4 W	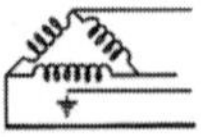	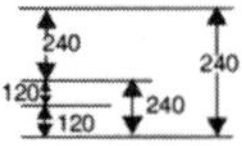 240 240 120 / 120 240	Motor and lighting loads

FIGURE A.7 Characteristics of electrical systems.

Trade Name	Type Letter	Temperature Rating	Application Provisions
Rubber-covered fixture wire (solid or 7-strand)	*RF-1	60°C 140°F	Fixture wiring, limited to 300 V
	*RF-2	60°C 140°F	Fixture wiring
Rubber-covered fixture wire (flexible stranding)	*FF-1	60°C 140°F	Fixture wiring limited to 300 V
	*FF-2	60°C 140°F	Fixture wiring
Heat-resistant, rubber-covered fixture wire (solid or 7-strand)	*RFH-1	75°C 167°F	Fixture wiring, limited to 300 V
	*RFH-2	75°C 167°F	Fixture wiring
Heat-resistant, rubber-covered fixture wire (flexible stranding)	*FFH-1	75°C 167°F	Fixture wiring, limited to 300 V
	*FFH-2	75°C 167°F	Fixture wiring
Thermoplastic-covered fixture wire (solid or stranded)	*TF	60°C 140°F	Fixture wiring
Thermoplastic-covered fixture wire (flexible stranding)	*TFF	60°C 140°F	Fixture wiring
Cotton-covered, heat-resistant fixture wire	*CF	90°C 194°F	Fixture wiring, limited to 300 V
Asbestos-covered, heat-resistant fixture wire	*AF	150°C 302°F	Fixture wiring, limited to 300 V and dry indoor locations
Silicone, rubber-insulated fixture wire (solid or 7-strand)	*SF-1	200°C 392°F	Fixture wiring, limited to 300 V
	*SF-2	200°C 392°F	Fixture wiring
Silicone, rubber-insulated fixture wire (flexible stranding)	*SFF-1	150°C 302°F	Fixture wiring, limited to 300 V
	*SFF-2	150°C 302°F	Fixture wiring
Code rubber	R	60°C 140°F	Dry locations
Heat-resistant rubber	RH	75°C 167°F	Dry locations
	RHH	90°C 194°F	Dry locations

FIGURE A.8a Conductor insulation.

Trade Name	Type Letter	Temperature Rating	Application Provisions
Moisture-resistant rubber	RW	60°C 140°F	Dry and wet locations; for over 2,000 V, insulation shall be ozone-resistant.
Moisture- and heat-resistant rubber	RH-RW	60°C 140°F	Dry and wet locations; for over 2,000 V, insulation shall be ozone-resistant.
		75°C 167°F	Dry locations
Thermoplastic and fibrous outer braid	TBS	90°C 194°F	Switchboard wiring only
Synthetic heat-resistant	SIS	90°C 194°F	Switchboard wiring only
Mineral insulation (metal sheathed)	MI	85°C 185°F	Dry and wet locations with Type O termination fittings, maximum operating temperature for special applications 250°C
Silicone-asbestos	SA	90°C 194°F	Dry locations, maximum operating temperature for special applications 125°C
Fluorinated ethylene propylene	FEP	90°C 194°F	Dry locations
	FEPB	200°C 392°F	Dry locations—special applications
		75°C 167°F	For over 2,000 V, insulation shall be ozone-resistant.
Moisture- and heat-resistant rubber	RHW	75°C 167°F	Dry and wet locations; for over 2,000 V, insulation shall be ozone-resistant
Latex rubber	RU	60°C 140°F	Dry locations
Heat-resistant latex rubber	RUH	75°C 167°F	Dry locations
Moisture-resistant latex rubber	RUW	60°C 140°F	Dry and wet locations
Thermoplastic	T	60°C 140°F	Dry locations
Moisture-resistant thermoplastic	TW	60°C 140°F	Dry and wet locations
Heat-resistant thermoplastic	THHN	90°C 194°F	Dry locations
Moisture- and heat-resistant thermoplastic	THW	75°C 167°F	Dry and wet locations
	THWN	75°C 167°F	Dry and wet locations

FIGURE A.8b Conductor insulation.

Size	Temperature Rating of Conductor[1]					
AWG or MCM	60°C (140°F)	75°C (167°F)	85-90°C (185°F)	110°C (230°F)	125°C (257°F)	200°C (392°F)
14	15	15	25[2]	30	30	30
12	20	20	30[2]	35	40	40
10	30	30	40[2]	45	50	55
8	40	45	50	60	65	70
6	55	65	70	80	85	95
4	70	85	90	105	115	120
3	80	100	105	120	130	145
2	95	115	120	135	145	165
1	110	130	140	160	170	190
1/0	125	150	155	190	200	225
2/0	145	175	185	215	230	250
3/0	165	200	210	245	265	285
4/0	195	230	235	275	310	340
250	215	255	270	315	335	—
300	240	285	300	345	380	—
350	260	310	325	390	420	—
400	280	335	360	420	450	—
500	320	380	405	470	500	—
600	355	420	455	525	545	—
700	385	460	490	560	600	—
750	400	475	500	580	620	—
800	410	490	515	600	640	—
900	435	520	555	—	—	—
1,000	455	545	585	680	730	—
1,250	495	590	645	—	—	—
1,500	520	625	700	785	—	—
1,750	545	650	735	—	—	—
2,000	560	665	775	840	—	—
°C / °F	Correction Factors, Room Temperatures Over 30°C (86°F)					
40 / 104	0.82	0.88	0.90	0.94	0.95	—
45 / 113	0.71	0.82	0.85	0.90	0.92	—
50 / 122	0.58	0.75	0.80	0.87	0.89	—
55 / 131	0.41	0.67	0.74	0.83	0.86	—
60 / 140	—	0.58	0.67	0.79	0.83	0.91
70 / 158	—	0.35	0.52	0.71	0.76	0.87
75 / 167	—	—	0.43	0.66	0.72	0.86
80 / 176	—	—	0.30	0.61	0.69	0.84
90 / 194	—	—	—	0.50	0.61	0.80
100 / 212	—	—	—	—	0.51	0.77
120 / 248	—	—	—	—	—	0.69
140 / 284	—	—	—	—	—	0.50

[1]Current capacities relate to conductors in Table B-2. See Table B-2 for the temperature rating of the conductor.
[2]Current capacities for types FEP, FEPB, PHH, and THHN conductors for sizes AWG 14, 12, and 10 shall be the same as designated for 75°C conductors in this table.

FIGURE A.9 Allowable current-carrying capacity of copper conductors—not more than three conductors in raceway or cable.

Size	Temperature Rating of Conductor[1]						
AWG or MCM	60°C (140°F)	75°C (167°F)	85-90°C (185°F)	110°C (230°F)	125°C (257°F)	200°C (392°F)	Bare and Covered Conductor
14	20	20	30[2]	40	40	45	30
12	25	25	40[2]	50	50	55	40
10	40	40	55[2]	65	70	75	55
8	55	65	70	85	90	100	70
6	80	95	100	120	125	135	100
4	105	125	135	160	170	180	130
3	120	145	155	180	195	210	150
2	140	170	180	210	225	240	175
1	165	195	210	245	265	280	205
1/0	195	230	245	285	305	325	235
2/0	225	265	285	330	355	370	275
3/0	260	310	330	385	410	430	320
4/0	300	360	385	445	475	510	370
250	340	405	425	495	530	—	410
300	375	445	480	555	590	—	460
350	420	505	530	610	655	—	510
400	455	545	575	665	710	—	555
500	515	620	660	765	815	—	630
600	575	690	740	855	910	—	710
700	630	755	815	940	1,005	—	780
750	655	785	845	980	1,045	—	810
800	680	815	880	1,020	1,085	—	845
900	730	870	940	—	—	—	905
1,000	780	935	1,000	1,165	1,240	—	965
1,250	890	1,065	1,130	—	—	—	—
1,500	980	1,175	1,260	1,450	—	—	1,215
1,750	1,070	1,280	1,370	—	—	—	—
2,000	1,155	1,385	1,470	1,715	—	—	1,405
°C	°F	Correction Factors, Room Temperatures Over 30°C (86°F)					
40	104	0.82	0.88	0.90	0.94	0.95	—	—
45	113	0.71	0.82	0.85	0.90	0.92	—	—
50	122	0.58	0.75	0.80	0.87	0.89	—	—
55	131	0.41	0.67	0.74	0.83	0.86	—	—
60	140	—	0.58	0.67	0.79	0.83	0.91	—
70	158	—	0.35	0.52	0.71	0.76	0.87	—
75	167	—	—	0.43	0.66	0.72	0.86	—
80	176	—	—	0.30	0.61	0.69	0.84	—
90	194	—	—	—	0.50	0.61	0.80	—
100	212	—	—	—	—	0.51	0.77	—
120	248	—	—	—	—	—	0.69	—
140	284	—	—	—	—	—	0.50	—

[1]Current capacities relate to conductors in Table B-2, pages B-2 through B-4. See Table B-2 for the temperature rating of the conductor.
[2]Current capacities for types FEP, FEPB, PHH, and THHN conductors for sizes AWG 14, 12, and 10 shall be the same as designated for 75°C conductors in this table.

FIGURE A.10 Allowable current-carrying capacity of copper conductors in free air.

Size	Temperature Rating of Conductor[1]					
AWG or MCM	60°C (140°F)	75°C (167°F)	85-90°C (185°F)	110°C (230°F)	125°C (257°F)	200°C (392°F)
12	15	15	25[2]	30	30	30
10	25	25	30[2]	35	40	45
8	30	40	40[2]	45	50	55
6	40	50	55	60	65	75
4	55	65	70	80	90	95
3	65	75	80	95	100	115
2[3]	75	90	95	105	115	130
1[3]	85	100	110	125	135	150
1/0[3]	100	120	125	150	160	180
2/0[3]	115	135	145	170	180	200
3/0[3]	130	155	165	195	210	225
4/0[3]	155	180	185	215	245	270
250	170	205	215	250	270	190
300	190	230	240	275	305	—
350	210	250	260	310	335	—
400	225	270	290	335	360	—
500	260	310	330	380	405	—
600	285	340	370	425	440	—
700	310	375	395	455	485	—
750	320	385	405	470	500	—
800	330	395	415	485	520	—
900	355	425	455	—	—	—
1,000	375	445	480	560	600	—
1,250	405	485	530	—	—	—
1,500	435	520	580	650	—	—
1,750	455	545	615	—	—	—
2,000	470	560	650	705	—	—

°C	°F	Correction Factors, Room Temperatures Over 30°C (86°F)				
40	104	0.82	0.88	0.90	0.94	0.95	—
45	113	0.71	0.82	0.85	0.90	0.92	—
50	122	0.58	0.75	0.80	0.87	0.89	—
55	131	0.41	0.67	0.74	0.83	0.86	—
60	140	—	0.58	0.67	0.79	0.83	0.91
70	158	—	0.35	0.52	0.71	0.76	0.87
75	167	—	—	0.43	0.66	0.72	0.86
80	176	—	—	0.30	0.61	0.69	0.84
90	194	—	—	—	0.50	0.61	0.80
100	212	—	—	—	—	0.51	0.77
120	248	—	—	—	—	—	0.69
140	284	—	—	—	—	—	0.50

[1]Current capacities relate to conductors in Table B-2, pages B-2 through B-4. See Table B-2 for the temperature rating of the conductor.

[2]Current capacities for types FEP, FEPB, PHH, and THHN conductors for sizes AWG 14, 12, and 10 shall be the same as designated for 75°C conductors in this table.

[3]For three-wire, single-phase service and subservice circuits, the allowable current capacity of RH, RH-RW, RHH, RHW, and THW aluminum conductors shall be for sizes No 2-100 amp, No 1-110 amp, No 1/0-125 amp, No 2/0-50 amp, No 3/0-170 amp, and No 4/0-200 amp.

FIGURE A.11 Allowable current-carrying capacity of aluminum conductors—not more than three conductors in raceway or cable.

Size	Temperature Rating of Conductor[1]						
AWG or MCM	60°C (140°F)	75°C (167°F)	85-90°C (185°F)	110°C (230°F)	125°C (257°F)	200°C (392°F)	Bare and Covered Conductor
12	20	20	30[2]	40	40	45	30
10	30	30	45[2]	50	55	60	45
8	45	55	55[2]	65	70	80	55
6	60	75	80	95	100	105	80
4	80	100	105	125	135	140	100
3	95	115	120	140	150	165	115
2	110	135	140	165	175	185	135
1	130	155	165	190	205	220	160
1/0	150	180	190	220	240	255	185
2/0	175	210	220	255	275	290	215
3/0	200	240	255	300	320	335	250
4/0	230	280	300	345	370	400	290
250	265	315	330	385	415	—	320
300	290	350	375	435	460	—	360
350	330	395	415	475	510	—	400
400	355	425	450	520	555	—	435
500	405	485	515	595	635	—	490
600	455	545	585	675	720	—	560
700	500	595	645	745	795	—	615
750	515	620	670	775	825	—	640
800	535	645	695	805	855	—	670
900	580	700	750	—	—	—	725
1,000	625	750	800	930	990	—	770
1,250	710	855	905	—	—	—	—
1,500	795	950	1,020	1,175	—	—	985
1,750	875	1,050	1,125	—	—	—	—
2,000	960	1,150	1,220	1,425	—	—	1,165

°C	°F	Correction Factors, Room Temperatures Over 30°C (86°F)					
40	104	0.82	0.88	0.90	0.94	0.95	—	—
45	113	0.71	0.82	0.85	0.90	0.92	—	—
50	122	0.58	0.75	0.80	0.87	0.89	—	—
55	131	0.41	0.67	0.74	0.83	0.86	—	—
60	140	—	0.58	0.67	0.79	0.83	0.91	—
70	158	—	0.35	0.52	0.71	0.76	0.87	—
75	167	—	—	0.43	0.66	0.72	0.86	—
80	176	—	—	0.30	0.61	0.69	0.84	—
90	194	—	—	—	0.50	0.61	0.80	—
100	212	—	—	—	—	0.51	0.77	—
120	248	—	—	—	—	—	0.69	—
140	284	—	—	—	—	—	0.50	—

[1]Current capacities relate to conductors in Table B-2, pages B-2 through B-4. See Table B-2 for the temperature rating of the conductor.
[2]Current capacities for types FEP, FEPB, PHH, and THHN conductors for sizes AWG 14, 12, and 10 shall be the same as designated for 75°C conductors in this table.

FIGURE A.12 Allowable current-carrying capacity of aluminum conductors in free air.

Occupancy	Standard Loads, Watts per square foot	Feeder-Demand Factor
ories and auditoriums	1	100%
ks	2	100%
er shops	3	100%
ches	1	100%
s	2	100%
lings	3	100% for first 3,000 watts, 35% for next 117,000, 25% for excess above 120,000
ges	0.5	100%
itals	2	40% for first 50,000 watts, 20% for excess over 50,000
e buildings	5	100% for first 20,000 watts, 70% for excess over 20,000
aurants	2	100%
ols	3	100% for first 15,000 watts, 50% for excess over 15,000
es	3	100%
houses	0.25	100% for first 12,500 watts, 50% for excess over 12,500
mbly halls	1	100%

RE A.14 Standard loads for branch circuits and feeders and demand
s for feeders.

...FEP, FEPB, R, RW, RU, RUW, RH-RW, SA, T, TW, RH, RUH, RHW, RHH, THHN, THW, and THWN Conductors in Raceway or Cable					
Rating	15 amp	20 amp	30 amp	40 amp	50 amp
ctors (min size): it wires*	14 / 14	12 / 14	10 / 14	8 / 12	6 / 12
urrent protection	15 amp	20 amp	30 amp	40 amp	50 amp
devices: holders permitted ptacle rating	Any type 15 max amp	Any type 15 or 20 amp	Heavy duty 30 amp	Heavy duty 40 and 50 amp	Heavy duty 50 amp
um load	15 amp	20 amp	30 amp	40 amp	50 amp

current capacities are for copper conductors with no correction factor for temperature (see Tables B-3
B-7, pages B-5 through B-9).

RE A.15 Requirements for branch circuits.

Trade Name	Type Letter	AWG	No of Conductors	Insulation	Braid on Each Conductor	Outer Covering	Use		
Parallel tinsel cord	TP	27	2	Rubber	None	Rubber	Attached to an appliance	Damp places	Not hard usage
	TPT	27	2	Thermoplastic	None	Thermoplastic	Attached to an appliance	Damp places	Not hard usage
Jacketed tinsel cord	TS	27	2 or 3	Rubber	None	Rubber	Attached to an appliance	Damp places	Not hard usage
	TST	27	2 or 3	Thermoplastic	None	Thermoplastic	Attached to an appliance	Damp places	Not hard usage
Asbestos-covered, heat-resistant cord	AFC	18-10	2 or 3	Impregnated asbestos	Cotton or rayon	None	Pendant	Dry places	Not hard usage
	AFPO	18-10	2	Impregnated asbestos	Cotton or rayon	Cotton, rayon, or saturated asbestos	Pendant	Dry places	Not hard usage
	AFPD	18-10	2 or 3	Impregnated asbestos	None	Cotton, rayon, or saturated asbestos	Pendant	Dry places	Not hard usage
Cotton-covered, heat-resistant cord	CFC	18-10	2 or 3	Impregnated cotton	Cotton or rayon	None	Pendant	Dry places	Not hard usage
	CFPO	18-10	2	Impregnated cotton	Cotton or rayon	Cotton or rayon	Pendant	Dry places	Not hard usage
	CFPD	18-10	2 or 3	Impregnated cotton	None	Cotton or rayon	Pendant	Dry places	Not hard usage
Parallel cord	PO-1	18	2	Rubber	Cotton	Cotton or rayon	Pendant or portable	Dry places	Not hard usage
	PO-2	18-16	2	Rubber	Cotton	Cotton or rayon	Pendant or portable	Dry places	Not hard usage
	PO	18-10	2	Rubber	Cotton	Cotton or rayon	Pendant or portable	Dry places	Not hard usage
All-rubber parallel cord	SP-1	18	2	Rubber	None	Rubber	Pendant or portable	Damp places	Not hard usage
	SP-2	18-16	2	Rubber	None	Rubber	Pendant or portable	Damp places	Not hard usage
	SP-3	18-10	2	Rubber	None	Rubber	Refrigerators or room air conditioners	Damp places	Not hard usage

FIGURE A.13a Flexible cords.

Trade Name	Type Letter	AWG	No of Conductors	Insulation	Braid on Each Conductor	Outer Covering	Use		
All-plastic parallel cord	SPT-1	18	2	Thermoplastic	None	Thermoplastic	Pendant or portable	Damp places	Not hard usage
	SPT-2	18-16	2	Thermoplastic	None	Thermoplastic	Pendant or portable	Damp places	Not hard usage
	SPT-3	18-10	2	Thermoplastic	None	Thermoplastic	Refrigerators or room air conditioners	Damp places	Not hard usage
Lamp cord	C	18-10	2 or more	Rubber	Cotton	None	Pendant or portable	Dry places	Not hard usage
Twisted, portable cord	PD	18-10	2 or more	Rubber	Cotton	Cotton or rayon	Pendant or portable	Dry places	Not hard usage
Reinforced cord	P-1	18	2 or more	Rubber	Cotton	Cotton over rubber filler	Pendant or portable	Dry places	Not hard usage
	P-2	18-16	2 or more	Rubber	Cotton	Cotton over rubber filler	Pendant or portable	Dry places	Not hard usage
	P	18-10	2 or more	Rubber	Cotton	Cotton over rubber filler	Pendant or portable	Dry places	Hard usage
Braided, heavy-duty cord	K	18-10	2 or more	Rubber	Cotton	Two-cotton, moisture-resistant finish	Pendant or portable	Damp places	Hard usage
Vacuum-cleaner cord	SV, SVO	18	2	Rubber	None	Rubber	Pendant or portable	Damp places	Not hard usage
	SVT, SVTO	18	2	Thermoplastic	None	Thermoplastic	Pendant or portable	Damp places	Not hard usage
Junior hard-service cord	SJ	18-16	2, 3 or 4	Rubber	None	Rubber	Pendant or portable	Damp places	Hard usage
	SJO	18-16	2, 3 or 4	Rubber	None	Oil-resistant compound	Pendant or portable	Damp places	Hard usage
	SJT, SJTO	18-16	2, 3 or 4	Thermoplastic or rubber	None	Thermoplastic	Pendant or portable	Damp places	Hard usage
Hard-service cord	S	18-2	2 or more	Rubber	None	Rubber	Pendant or portable	Damp places	Extra hard usage
	SO	18-2	2 or more	Rubber	None	Oil-resistant compound	Pendant or portable	Damp places	Extra hard usage
	ST	18-2	2 or more	Thermoplastic or rubber	None	Thermoplastic	Pendant or portable	Damp places	Extra hard usage
	STO	18-2	2 or more	Thermoplastic or rubber	None	Oil-resistant thermoplastic	Pendant or portable	Damp places	Extra hard usage

FIGURE A.13b   Flexible cords.

Trade Name	Type Letter	AWG	No of Conductors	Insulation	Braid on Each Conductor	Outer Covering	
Rubber-jacketed, heat-resistant cord	AFSJ	18-16	2 or 3	Impregnated asbestos	None	Rubber	Portable
	AFS	18-16-14	2 or 3	Impregnated asbestos	None	Rubber	Portable
Heater cord	HC	18-12	2, 3, or 4	Rubber and asbestos	Cotton	None	Portable
	HPD	18-12	2, 3, or 4	Rubber with asbestos or all neoprene	None	Cotton or rayon	Portable
Rubber-jacketed heater cord	HSJ	18-16	2, 3, or 4	Rubber with asbestos or all neoprene	None	Cotton and rubber	Portable
Jacketed heater cord	HSJO	18-16	2, 3, or 4	Rubber with asbestos or all neoprene	None	Cotton and oil-resistant compound	Portable
	HS	14-12	2, 3, or 4	Rubber with asbestos or all neoprene	None	Cotton and rubber or neoprene	Portable
	HSO	14-12	2, 3, or 4	Rubber with asbestos or all neoprene	None	Cotton and oil-resistant compound	Portable
Parallel heater cord	HPN	18-16	2	Thermosetting	None	Thermosetting	Portable
Heat- and moisture-resistant cord	AVPO	18-10	2	Asbestos and varnished cambric	None	Asbestos, flame-retardant, moisture-resistant	Pendant portable
	AVPD	18-10	2 or 3	Asbestos and varnished cambric	None	Asbestos, flame-retardant, moisture-resistant	Pendant portable
Range, dryer cable	SRD	10-4	3 or 4	Rubber	None	Rubber or neoprene	Portable
	SRDT	10-4	3 or 4	Thermoplastic	None	Thermoplastic	Portable
Elevator cable	E	18-14	2 or more	Rubber	Cotton	Three cotton, outer one flame-retardant and moisture-resistant	Elevator lighting control
	EO	18-14	2 or more	Rubber	Cotton	One cotton and a neoprene jacket	Elevator lighting control
	ET	18-14	2 or more	Thermoplastic	Rayon	Three cotton, outer one flame-retardant and moisture-resistant	Elevator lighting control

FIGURE A.13c   Flexible cords.

INDEX